STARRY WONDERS

STARRY WONDERS

Exploring the 25 Brightest Stars

Janine Bonham
Illustrated by Erin Miller

MIAMI

Copyright © 2025 by Janine Bonham.
Published by Mango Publishing, a division of Mango Publishing Group, Inc.

Cover Design: Elina Diaz
Cover Photo: stock.adobe.com/sripfoto
Interior Illustrations: Erin Miller
Interior Photography: stock.adobe.com
Layout & Design: Elina Diaz

Mango is an active supporter of authors' rights to free speech and artistic expression in their books. The purpose of copyright is to encourage authors to produce exceptional works that enrich our culture and our open society.

Uploading or distributing photos, scans or any content from this book without prior permission is theft of the author's intellectual property. Please honor the author's work as you would your own. Thank you in advance for respecting our author's rights.

For permission requests, please contact the publisher at:
Mango Publishing Group
5966 South Dixie Highway, Suite 300
Miami, FL 33143
info@mango.bz

For special orders, quantity sales, course adoptions and corporate sales, please email the publisher at sales@mango.bz. For trade and wholesale sales, please contact Ingram Publisher Services at customer.service@ingramcontent.com or +1.800.509.4887.

Starry Wonders: Exploring the 25 Brightest Stars

Library of Congress Cataloging-in-Publication number: 2024947351
ISBN: (print) 978-1-68481-475-6, (ebook) 978-1-68481-477-0
BISAC category code: SCI098030, SCIENCE / Space Science / Stellar & Solar

To my parents,
who wish as much as I do,
to go boldly where no one has gone before.

FOR MY PART, I KNOW NOTHING WITH ANY CERTAINTY, BUT THE SIGHT OF THE STARS MAKES ME DREAM.

—VINCENT VAN GOGH

TABLE OF CONTENTS

Preface	12
Stargazing	15
Ancient Beginnings	16
Modern Studies	17
Stargazing Basics	22
Constellations	25
Asterisms	29
Bortle Night Sky Classification	32
International Dark-Sky Parks	35
Celestial Objects	38
Celestial Events	50
Stars	57
Birth, Life, and Death of Stars	58
Types of Stars	64
Origin of Star Names	66
The Sun—Our Closest Star	68
25 Brightest Stars in the Night Sky	78
Sirius (Canis Major)—The Brightest Star in the Night Sky	79
Canopus (Carina)—Old Man of the South Pole	83
Rigil Kentaurus (Centaurus)—Our Closest Neighbor	86
Arcturus (Boötes)—Guardian of the Bear	89
Vega (Lyra)—Jewel of the Lyre	93
Capella (Auriga)—The Goat Star	101
Rigel (Orion)—Foot of the Giant	105
Procyon (Canis Minor)—The Lesser Dog Star	109
Achernar (Eridanus)—End of the River	113
Betelgeuse (Orion)—Red Supergiant in the Hunter	116
Hadar (Centaurus)—Triple Star System	120
Altair (Aquila)—Eye of the Eagle	123
Acrux (Crux)—Brightest of the Southern Cross	128
Aldebaran (Taurus)—The Follower	132
Antares (Scorpius)—Heart of the Scorpion	136

Spica (Virgo)—Ear of Wheat Star	140
Pollux (Gemini)—Red Giant of Gemini	143
Fomalhaut (Piscis Austrinus)—The Loneliest Star	147
Deneb (Cygnus)—Tail of the Swan	152
Mimosa (Crux)—Binary Star System	158
Regulus (Leo)—Heart of the Lion	161
Adhara (Canis Major)—Binary Star System	165
Shaula (Scorpius)—The Scorpion's Stinger	168
Castor (Gemini)—Six Star System	171
Gacrux (Crux)—The Closest Giant Star	174
Unique Stars	**185**
Closest Star: Proxima Centauri	186
Smallest: EBLM J0555-57Ab (57 Ab, for short)	187
RMC 136a1: The Most Massive Star Yet Known	188
Mira: A Wonderful Star with a Tail	189
Lucy (in the Sky with Diamonds)	190
Algol: The Demon Star	191
Notable Star Clusters	**195**
Pleiades	196
Hyades	199
Beehive Cluster	201
Coma Cluster	202
Alpha Persei Cluster	203
Hercules Cluster	204
Southern Pleiades	207
FAQ	**208**
Appendix	**216**
Glossary	**220**
References	**227**
Photo Credits	**231**
Acknowledgements	**236**
About the Author	**238**
About the Illustrator	**239**

PREFACE

I will never forget the first time I witnessed a night sky free from light pollution. It was my inaugural trip to Hawai'i, where I was participating in an intensive three-week college course that focused on the geological evolution of the Hawaiian Islands. One evening, a few classmates and I ventured to the Mauna Kea Visitor Information Station, and that night changed my life forever. I was awestruck by the sheer number of stars visible, and I found myself genuinely perplexed by the hazy band stretching across the sky—it was my first glimpse of the Milky Way galaxy. At that moment, I realized I would return to this enchanting place again and again.

Little did I know that in the years to come, I would end up living in Hawai'i, where I would teach and immerse myself in the sciences I cherish: astronomy, biology, marine science, and volcanology. It was in this paradise that I developed my skills as a stargazer. During my twelve years in Hawai'i, beneath exceptionally dark skies, I began to recognize the constellations and learned when they would be visible throughout the year. I spent countless nights outside my home, lying on blankets with my daughters, marveling at the stunning night sky. When my girls drifted off to sleep, I would pull out my field guide to identify what I was seeing. One of my favorite pastimes was using my binoculars to explore seemingly empty patches of sky, only to uncover hidden treasures of distant starlight.

As an educator and enthusiastic stargazer, I believe everyone should have the opportunity to witness the night sky free from artificial light. However, as time goes on, experiencing a truly dark sky has become increasingly challenging for most people. A significant portion of the world's population resides in urban and suburban areas, which hinders their ability to see what our ancient ancestors once viewed. Yet, even in places plagued by light pollution, some stars remain visible. It was these brightest stars that first captured my interest. What are these stars like? Are they part of binary systems? Do they have planets orbiting them? Why do they shine so brightly? Even after writing this book, I still find myself filled with questions.

My goal in writing this book was to highlight the celestial systems that are the brightest in our sky. I wanted to spotlight not only the most brilliant stars but also the intricate systems that characterize these celestial bodies. I encourage you to begin with the basics of stargazing to guide you through the process of having enjoyable and rewarding stargazing sessions. In the subsequent chapters, I delve into the life cycle of a star and how stars are classified. Understanding this information will enhance your knowledge of the twenty-five brightest stars and reveal their similarities and differences with our sun and one another. Finally, we will examine some truly unique stars that exist in our universe and notable star clusters that we can see with the unaided eye.

I also wanted to make the night sky more approachable for those who might feel intimidated by the complexities of astronomy. By highlighting the brightest and most recognizable celestial objects, my goal is to create an entry point for beginner stargazers to connect with the universe. I believe that by understanding these key systems, readers will gain the confidence to venture outside, look up, and appreciate the beauty that has inspired humanity for centuries. Ultimately, my hope is that this book not only serves as a guide to the brightest points of light in our night sky but also ignites a passion for exploration and a deeper appreciation for the wonders of the cosmos.

STAR GAZING

Ancient Beginnings

Humanity has been gazing up at the night sky for thousands of years. It is the one thing we all share together.

Some of the earliest human artifacts that have been discovered depict the arrangements of specific star shapes and patterns. The apparent movement of the sun, moon, stars, and planets across the sky makes a giant celestial clock. The earliest civilizations recorded these movements to help time agricultural activities, mark the changing of the seasons, and guide travelers. The stories of the stars were often meaningful and helped to preserve cultural legacy. Humans have often associated these star patterns with mythical people and religious themes throughout history. Legends from around the world reveal ancient knowledge of the sky, and these stories have been passed down from generation to generation. Mythology, religion, and cultural practices were frequently interwoven in ancient astronomy.

In the past, astronomers could only view the sky with their unaided eyes. To follow the motion of the planets and stars, they employed basic sighting devices and gnomons, which are sticks that produce shadows. Observations were documented using rudimentary means such as carvings, writing on clay tablets, and drawings. This data was used to track cosmic patterns throughout time.

Most ancient civilizations, including the Greeks, Egyptians, and Babylonians, believed in a geocentric model of the universe, where Earth was at the center and everything else orbited around it. Celestial phenomena were often explained through myths and legends and lacked a scientific basis. Figures such as Ptolemy, who formulated the geocentric model using epicycles, and Hipparchus, who cataloged stars and discovered precession, made significant contributions that shaped our understanding of the cosmos.

The Mayans, Chinese, and Indians also made significant contributions with their unique observations and calendars. Although the myths and legends of the stars differ depending on time, place, and culture, the curiosity about what is beyond Earth's borders is always there. Everyone shares the same sky.

Modern Studies

The study of astronomy has evolved significantly from ancient times to the modern era. The dawn of modern astronomy began with the start of the Renaissance period in the sixteenth century. Multiple figures in astronomy began to look at the cosmos in a different way. The invention of the telescope, the scientific method, and using mathematics to explain observations began to take hold in the astronomy community.

FIRST MODERN ASTRONOMERS

 **Nicolaus Copernicus (1473-1543)** Copernicus suggested a model of the solar system in which the sun, not the earth, is at the center. His 1543 book, *De revolutionibus orbium coelestium* (On the Revolutions of the Celestial Spheres), is often seen as the beginning of modern astronomy.

 **Tycho Brahe (1546-1601)** Brahe carefully and extensively observed the planets and stars, which later scientists used to make important discoveries. Tycho built two telescopes called Uraniborg and Stjerneborg. These telescopes had big, accurate tools for making observations with just the naked eye.

 **Johannes Kepler (1571-1630)** Kepler used Brahe's data to come up with his three laws of planetary motion. These laws explained how planets move in elliptical pathways (not perfect circles) and set the stage for Newton's theory of gravity.

 **Galileo Galilei (1564-1642)** Galileo's use of the telescope to study the stars changed the field of astronomy forever. He found moons that circled Jupiter, watched Venus's phases, and looked into sunspots, all of which strongly supported the Copernican system.

 **Isaac Newton (1643-1727)** In his *Principia Mathematica* (1687) publication, Newton laid out a thorough framework for comprehending the movement of heavenly things, which included his law of universal gravity and his contributions to classical mechanics.

More discoveries concerning the distant universe have been made during the last two hundred years as technology has advanced. Advances in both ground-based and space-based observations continue to increase, allowing humans to see parts of the night sky that are otherwise undetectable to the naked eye. Modern astronomy is a scientific discipline focused on understanding the universe's fundamental laws and phenomena. It encompasses cosmology, the study of the universe's origin, evolution, and eventual fate. Astronomy contributes to advancements in technology, including satellite communications, GPS, and even medical imaging techniques.

Modern astronomers and ancient stargazers all possess the desire to learn the patterns and mysteries of the night sky. Ancient astronomy was restricted to what could be seen with the naked eye and was frequently geocentric. Modern astronomy uses advanced technologies to examine the cosmos at all scales, from subatomic particles to large cosmic structures. From ancient to modern times, astronomy has changed from a field-based observation and interpretation to a strict, empirical science that uses cutting-edge technology and theoretical theories to study the universe.

Contemporary astronomers study celestial objects in great detail using sophisticated telescopes. Modern telescopes are often specified to examine a celestial object in a particular wavelength. These include radio, microwaves, infrared, optical, ultraviolet, and X-rays. Telescopes are space-based and ground-based and the use of computers for data processing, simulation, and modeling are all essential components of modern-day astronomy. Observations coupled with data analysis enable more precise and complicated investigations. Modern research methods include photometry, which measures the brightness of distant objects, and spectroscopy, which examines light from astronomical objects to determine their composition and other characteristics.

Despite all of the advances made by the astronomical community, the glittering stars are sometimes unseen to the majority of us. As the number of people living in cities has grown, so have the cities and the bright lights that come with them. This is called "urban sprawl." Cities grow as their people do, and so does the infrastructure they need to support them, like buildings, streetlights, and other sources of artificial light. The growing web of lights is called light pollution and it continues to negatively impact observational practices.

The consequences are twofold: light pollution not only obscures the night sky and reduces our capacity to see stars and other celestial phenomena, but it also impacts natural ecosystems, human health, and energy usage. The widespread use of artificial lighting has resulted in the loss of completely dark skies in many regions of the world, making it more difficult for people to appreciate the grandeur of a starry night sky. As cities grow larger and brighter, the need for secluded regions for stargazing grows, and for many individuals, this means driving vast miles to catch a glimpse of what was once visible from their backyard.

This growth has produced a complicated relationship between progress and preservation as humanity's footprint on the world grows more visible, putting the beauty and quiet of a naturally dark night sky out of grasp for most people. The breathtaking night sky is a treasure worth preserving. It has sparked the imagination of countless individuals and encourages us to wonder about what lies beyond the limits of our home planet.

MODERN-DAY ASTRONOMICAL TOOLS

Binoculars and Telescope

Telescope for Astrophotography

Radio Telescope

Gemini Optical Telescope

Hubble Space Telescope

James Webb Space Telescope

Stargazing Basics

Stargazing takes time, patience, practice, and darkness.

In the past two hundred years, the experience of stargazing has become vastly different than our ancient ancestors. As the human population continues to grow, more artificial light bleeds into the atmosphere, masking the distant light from far-away stars. The key to learning the star patterns and seeing distant nebulae and galaxies is to first find dark skies. This will give you the opportunity to really see what's out there.

Stargazing Tips

Seek out dark skies. Stargazing is at its best away from light pollution. This includes both natural and artificial light. Stargaze when the moon is absent, and travel away from city lights to get the best view. To get the best stargazing experience, it's recommended to travel at least twenty to fifty miles (thirty-two to eighty kilometers) away from city lights. The exact distance depends on the size of the city and how much light pollution it produces. Here's a general guideline:

1. **Small Cities or Towns:** Travel at least twenty miles (thirty-two km) away to reduce the impact of light pollution.

2. **Medium-Sized Cities:** Aim to be thirty to fifty miles (forty-eight to eighty km) away for darker skies and clearer views.

3. **Large Cities or Urban Areas:** For optimal stargazing conditions, try to get over fifty miles (over eighty km) away to escape the glow of light pollution.

The goal is to reach a location where the Milky Way is clearly visible, and the sky is dark enough to see stars down to the horizon. Checking a light pollution map can help you find the nearest "dark sky" area.

Give your eyes time to adjust to the dark. To get the best view of the stars, avoid white light. It can take up to thirty minutes for your eyes to adjust to the night sky. Giving your eyes at least five to ten minutes of adjustment will change what you see.

Learn the basics. Whenever you learn something new, it is important to use as many resources as possible to start your journey. When learning constellations, start with a star chart so you can learn the star patterns.

Get comfortable. Many newbie stargazers make the mistake of stargazing while standing. Using your neck to look up will only make your stargazing session short. Find a way to get comfortable. Try a lawn chair, blanket, or hammock. Be creative, avoid straining your neck, and stay warm if the season is cold.

Know your directions. Learning the cardinal directions in your area can be helpful when trying to find the constellations. Find **due north** by using the Big Dipper.

Share your experience. Stargazing with a companion is a great way to connect to the sky. Whether it be a family member, friend, or pet, spending time outdoors and observing the beauty of the night sky is sure to enhance your experience.

Beginner Tools

- **Planisphere/star wheel**—A map of the sky is always helpful when learning the constellations.

- **Red flashlight** (avoid white light)—Red light helps preserve your night vision. Our eyes are less sensitive to red light, which means red light won't cause your pupils to contract as much as other colors would. This allows your eyes to stay adapted to the darkness, enabling you to continue seeing dim stars and celestial objects. Exposure to white light causes your pupils to contract, reducing their ability to detect faint objects in the dark. It can take up to twenty to thirty minutes for your eyes to fully readjust and regain night vision after being exposed to bright light.

- **Recliner**—Looking straight up at the sky will put a strain on your neck. Find a way to lie back and get comfortable so you can stargaze longer.

- **Blanket**—Weather conditions change with the seasons, and without the warmth of the sun, nights can get quite chilly. Bringing a blanket can help keep you comfortable and cozy as you stargaze.

- **Telescope/binoculars**—Having a telescope or a pair of binoculars is not essential when you first stargaze, but if you find this hobby enjoyable, investing in equipment can enhance your experience. It's always captivating to gaze into what seems like a completely dark part of the sky, only to discover countless hidden stars when viewed through magnification.

- **Field guide**—A field guide provides detailed information on stars, constellations, planets, and deep-sky objects, helping you identify what you're seeing in the night sky. It shows you which stars and constellations are visible during different times of the year, making it easier to plan your observations and know what to look for in each season.

Constellations

Constellations are patterns in the sky that appear to be grouped together and they have been recorded over thousands of years of human history. Their patterns and stories may have shifted over time, but their use has remained the same—an ancient time clock that helps humans predict the seasons, plan agricultural practices, and navigate across the land and sea. However, it is important to recognize that the patterns of the stars vary according to time, place, and culture.

As European explorers ventured to the southern hemisphere, they saw stars that were not visible in the northern hemisphere. This resulted in the identification of new constellations. In 1603, Johann Bayer, a German astronomer, produced *Uranometria*, the first comprehensive star atlas. Bayer developed a system for identifying stars within constellations using Greek and Latin letters, which is still in use today. In the eighteenth century, astronomers such as Nicolas-Louis de Lacaille cataloged numerous constellations in the southern hemisphere, enlarging the list of known constellations. The development of telescopes and other astronomy tools enabled more exact observations, resulting in a greater understanding of star locations and movements.

By the late nineteenth and early twentieth centuries, astronomers understood the importance of a uniform constellation system. The International Astronomical Union (IAU) was established in 1919 to promote worldwide cooperation in astronomy. In 1922, the IAU officially accepted eighty-eight constellations, defining their names and limits. This system contained both ancient constellations from Ptolemy's list and more recent ones discovered during the Age of Exploration. Modern constellations are utilized for astronomy research and navigation. They offer a standardized method for recognizing and locating celestial objects.

Types of Constellations

CIRCUMPOLAR CONSTELLATIONS

There is a spot in our sky directly above Earth's North Pole. It is called the North Celestial Pole. The constellations closest to the North Celestial Pole never set below the horizon, so they are constantly visible throughout the night all year long. In the northern hemisphere there are five main circumpolar constellations that are easily identified: Ursa Major, Ursa Minor, Cassiopeia, Cepheus, and Draco. A couple of others, such as the Lynx and Camelopardalis, are also considered circumpolar but are not as visible as the ones previously mentioned.

Ursa Major and Ursa Minor are both pictured in this photo. The Big Dipper is an asterism within the Ursa Major pattern, and it can be used to find Ursa Minor that contains Polaris, the North Star.

The word *circumpolar* can be broken down to help you remember its meaning. The *circum-* prefix is derived from the Latin word meaning "around," and the *polar-* part of the word is in reference to the pole star, or the North Star. So when you put those words together, *circumpolar*, it means "around the pole star."

ZODIACAL CONSTELLATIONS

Every year, our sun takes the same predictable path in the sky. The path of the sun in our sky is called the **ecliptic**. Any constellation that lands on the path of the ecliptic is considered a zodiacal constellation. Astronomers recognize thirteen zodiacal constellations.

But wait—you might be thinking—aren't there twelve signs of the zodiac? Yes, there are, but that is in the world of *astrology*. Four centuries ago, astronomy and astrology were considered the same thing, but in modern times, they are separate. Astrology today is defined as the study of the motion of the stars and planets and how it influences human affairs and the natural world, and it is considered by many as pseudoscience. Modern astronomy is the scientific study of celestial objects, space, and the universe as a whole. It encompasses the observation and analysis of phenomena that originate beyond Earth's atmosphere, including stars, planets, moons, galaxies, and other cosmic entities.

The constellations that the sun, moon, and planets pass through are the constellations of the zodiac. The earliest records of these patterns are the Babylonian Star Catalogs. These tablets cite seventy-one stars and constellations and their connections to ancient deities. The information about the stars included constellations that rise and set together, the path of the moon and planets, a solar calendar, and the positions of stars throughout the night. Over time, many cultures adopted these patterns and the stories connected to them over the centuries. Today, astronomers identify the zodiac constellations as one in which the sun passes through. They are identified as follows: Aries, Taurus, Gemini, Cancer, Leo, Virgo, Libra, Scorpius, Ophiuchus, Sagittarius, Capricornus, Aquarius, and Pisces.

Also, a familiar question that many ask is: When can I see my personal zodiac constellation in the sky? The zodiac sign is in reference to when the **sun** is in that constellation. For example, if you are born under a particular sign, that constellation is not visible at night; rather the sun is passing through it around that time of year. The constellation cannot be visible because it is present during the day, and the sun blocks out the backdrop of stars behind it. Typically, your zodiac sign peaks around six months after your actual birthday.

SEASONAL CONSTELLATIONS

Seasonal constellations change along with the seasons. The seasonal climate changes will be specific depending upon your location on Earth, but the "seasons" of the sky are predictable. All constellations can be visible for about six months, but often constellations peak during particular times of the year.

Asterisms

Asterisms are not true constellations, but rather patterns that naturally connect in the sky and are often the easiest stars and patterns to find. Using these alternative star patterns can help new stargazers orient themselves with stellar navigation and star recognition.

All constellations and asterisms are man-made concepts passed down from generation to generation. They connect the ancient past of humanity with the modern-day study of the universe beyond our Earth. As modern astronomy aims to learn about the structure of the universe, we continue to look back at the ancient observers to learn how the skies have changed over the millennia.

COMMON ASTERISMS

Big Dipper

These seven stars make up a portion of the larger constellation Ursa Major. It has many names across the world.

Little Dipper

This star pattern contains Polaris, the current North Star. Earth's axis is oriented toward this star, and it appears to not move in the sky. Polaris has many names across the world, and it marks the end of the handle of the dipper shape.

Summer Triangle

Formed by three bright stars from different constellations: Vega in Lyra, Deneb in Cygnus, and Altair in Aquila.

STARGAZING

Great Square of Pegasus

An asterism formed by four stars in the constellation Pegasus, marking the body of the mythical winged horse.

Winter Triangle

Pattern of three stars that form a triangle. Sirius in Canis Major, Betelgeuse in Orion, and Procyon in Canis Minor.

Winter Hexagon

Formed by six bright stars from different constellations: Sirius in Canis Major, Procyon in Canis Minor, Pollux in Gemini, Capella in Auriga, Aldebaran in Taurus, and Rigel in Orion.

Northern Cross in Cygnus

An asterism formed by the arrangement of stars in the constellation Cygnus. It resembles a cross, with Deneb marking the head.

Teapot in Sagittarius

Star pattern resembling a teapot. The spout and lid are outlined by the brighter stars.

Golden Gate of the Ecliptic

The Pleiades (right) and the Hyades (left, V-shaped pattern) star clusters in the constellation Taurus intersect with the apparent path of the sun across the celestial sphere. The planet Mars is close to the Pleiades in this picture. All planets will pass between these two clusters.

Earth's city lights.

Bortle Night Sky Classification

The night sky has truly transformed itself in the past two centuries. As the human population continues to grow and urban areas become more densely populated, artificial light has made it more difficult to see the stars and the collective light from our galaxy. With each new generation, more light is being generated and erasing the stars from our collective memory. Global efforts are underway to promote safe lighting choices to lessen the overall consequences of light pollution.

Astronomers and amateur stargazers use the **Bortle Dark-Sky Scale** as a tool to measure how dark the sky is at a certain location. It was developed by John E. Bortle in 2001 and divides the sky into nine categories according to brightness and object visibility. Class 1 marks a great dark-sky site, while Class 9 indicates a night sky with lots of artificial light pollution, such as an inner-city sky. The Bortle scale is a useful tool for astronomers, astrophotographers, and stargazers to choose optimal locations for observing celestial objects and capturing images of the night sky. In addition to the quality of the night sky, all stargazers must also monitor the local weather conditions, as these events can also impact night sky visibility.

BORTLE CLASSIFICATION:

- Class 1: Excellent dark-sky site
    - The darkest skies possible on Earth.
    - Faintest objects visible in the universe can be seen.
- Class 2: Typical truly dark site
    - Excellent dark-sky conditions.
    - The Milky Way is prominent.
- Class 3: Rural sky
    - Rural locations with good dark-sky conditions.
    - The Milky Way is still impressive.
- Class 4: Rural/suburban transition
    - Rural areas with some light pollution from nearby towns or cities.
    - The Milky Way is still visible but with reduced detail.
- Class 5: Suburban sky
    - Suburban skies with moderate light pollution.
    - The Milky Way is visible but faint.
- Class 6: Bright suburban sky
    - Significant light pollution from surrounding urban areas.
    - The Milky Way is usually not visible.
- Class 7: Suburban/urban transition
    - Substantial light pollution from the city.
    - Only the brightest stars and planets are visible.
- Class 8: City sky
    - Heavy light pollution from the city.
    - Only the moon, planets, and a few bright stars are visible.
- Class 9: Inner-city sky
    - The most light-polluted skies.
    - Only the moon and a few bright stars and planets are visible.

International Dark-Sky Parks

Dark-sky parks are areas around the world that are designated for stargazing by reducing artificial light.

International Dark-Sky Parks (IDSPs) are designated areas that meet strict requirements for preserving and protecting the quality of the night sky. These parks are recognized by the International Dark-Sky Association (IDA), a nonprofit organization dedicated to combating light pollution and promoting responsible outdoor lighting.

To achieve Dark-Sky Park status, an area must demonstrate exceptional commitment to dark-sky preservation. This involves implementing specific outdoor lighting policies, educating the public about the importance of dark skies, and taking measures to reduce light pollution.

These parks are not only places where astronomers can observe celestial objects with minimal light pollution, but they also serve as important hubs for public education about the value of preserving dark skies. Visitors to these parks often participate in astronomy programs, stargazing events, and other educational activities focused on appreciating and protecting the night sky.

For many who grow up under light-polluted skies, seeing a truly dark sky with thousands of stars can be a moving experience. Escaping the city lights is essential when trying to see the farthest parts of the universe.

WELL-KNOWN DARK-SKY PARKS:

- Grand Canyon-Parashant National Monument (USA):
    - Located near the Grand Canyon in Arizona, this park offers stunning views of the night sky. The absence of major urban centers nearby contributes to its excellent dark-sky conditions.

- Jasper National Park (Canada):
    - Situated in the Canadian Rockies, Jasper National Park became the world's largest Dark-Sky Preserve in 2011. The park hosts various astronomy programs and events.

- NamibRand Nature Reserve (Namibia):
    - NamibRand was the first International Dark-Sky Reserve in Africa. Its remote location in the Namib Desert provides exceptional views of the southern hemisphere's night sky.

- Exmoor National Park (United Kingdom):
    - Exmoor, in southwest England, was designated an International Dark-Sky Reserve in 2011. The park works to reduce light pollution and promote astronomy-related activities.

- Aoraki Mackenzie Dark-Sky Reserve (New Zealand):
    - Covering a large portion of New Zealand's South Island, this reserve is known for its pristine night skies. The reserve includes Aoraki Mount Cook National Park and the Mackenzie Basin.

- Natural Bridges National Monument (USA):

- Located in Utah, this national monument became the first International Dark-Sky Park certified by the IDA in 2007. The park is known for its natural stone bridges and arches.

+ Mont-Mégantic National Park (Canada):

    - Located in Québec, this park was the first International Dark-Sky Reserve in the world. Its ASTROLab educational center provides opportunities for stargazing and learning about astronomy.

Aoraki Mackenzie Dark-Sky Reserve (New Zealand)

Celestial Objects

There is so much more to see in the sky than just stars! The planets of our solar system dance across the sky with positions that do not follow the regular timing of the star patterns. Tiny patches of concentrated stars make up star clusters, groups of stars bound together by gravity. Fuzzy patches of light and cloudiness indicate nebulae, pockets of gas and dust that serve as the birthplace of stars. Many of these objects were visible to the ancients, but with modern technology, astronomers can peer deeper into the universe and detect the faintest and farthest objects known.

Planets

Over the course of thousands of years, ancient civilizations from around the globe recorded the movement of the stars and predicted the motion of the planets. Planets move differently in the sky than stars. When stars rise and set throughout the year, their position and time of rise and set are the same day after day, year after year. But planets have different cycles than the stars. Sometimes they move forward, sometimes backward in comparison to the stars, and other times they appear to hold their position.

You can learn a few techniques to help you determine the visual difference between a planet and a star. Some planets, like Venus, Mars, and Jupiter, shine brighter than the surrounding stars. They also stay in the path of the ecliptic which is the pass of the sun. Any star pattern that intersects the ecliptic is classified as a zodiac constellation. So learning to recognize the zodiac constellations is a great start to learning where the planets will be in the sky. The motion of a planet is different from that of a star. A star has a predictable rise and set time for each day of the year, whereas planets wander around the night sky in a separate cycle along the ecliptic.

To classify a celestial body as a planet, it must be:

- A body that orbits a star
- Massive enough for its gravity to make it round
- Big enough that its gravity has cleared out the objects in its orbital path

Only five of the eight planets in our local solar system are visible to the unaided eye. These five were identified by the ancients, while the planets that are invisible to the eye were only discovered recently by modern-day astronomers. The planets Uranus and Neptune were discovered by scientists using telescopes.

Planet Position	Magnitude (Brightness)	Viewing Characteristics	Planet Type
Venus 2nd Planet	-4.7	Brightest planet. Often called the Morning Star or Evening Star. Best seen during sunrise and sunset. Can be seen during transits.	Terrestrial
Mars 4th Planet	-2.7	Reddish orange in color.	Terrestrial
Jupiter 5th Planet	-2.2	Four largest moons can be seen with a telescope. Banded planet.	Gas Giant
Saturn 6th Planet	-0.24	Rings can be visible and obvious when viewed through a telescope. Moves slowly in the sky compared to all other visible planets.	Gas Giant
Mercury 1st Planet	-2 to 7	Always close to the horizon during sunset or sunrise. Dim and difficult to see. Can be seen during transits.	Terrestrial
Uranus 7th Planet	5.6	Only planet named after a Greek god (the rest are Roman). Brightness near the naked eye limit.	Gas Giant
Neptune 8th Planet	7.7	Below the naked eye limit threshold.	Gas Giant

Nebula

A nebula (plural: nebulae, pronounced NE-byoo-lee) is a vast cloud of gas and dust in space. These clouds can vary in size from being relatively small, localized regions to enormous, diffuse structures that span many light-years. Nebulae are important in the field of astronomy because they often serve as the birthplaces of stars and pre-planetary systems. Nebulae are fascinating objects to study in astronomy because they provide insights into the processes of star and planet formation, the life cycles of stars, and the distribution of matter in the universe. They are often observed in various wavelengths of light, including visible, infrared, and radio, to reveal different aspects of their structure and composition.

STAR BIRTH

Emission Nebulae: These are clouds of gas (usually hydrogen) that emit light of various colors. They are often regions of active star formation, and the energy from young, hot stars embedded within the nebula causes the gas to glow.

Reflection Nebulae: These are similar to emission nebulae, but they don't emit their own light. Instead, they reflect the light from nearby stars, which illuminates the dust and gas within the nebula. The Pleiades Star cluster is perhaps the best and most well-known example of a reflection nebula. In this system, the clouds of gas and dust reflect the light from the nearby stars.

Dark Nebulae: Dark nebulae are cold, dense regions of gas and dust that block the light from stars and other objects behind them. They often appear as dark patches against a background of stars or other bright nebulae.

STAR DEATH

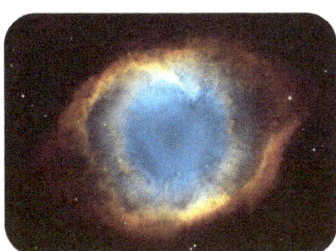

Planetary Nebulae: These are shells of gas and dust ejected by certain types of stars in the late stages of their evolution. Despite the name, they have nothing to do with planets. Instead, they are remnants of stars that have shed their outer layers. Planetary nebulae have a relatively brief lifetime in astronomical terms, lasting only a few tens of thousands of years before dispersing into the surrounding interstellar medium.

Supernova Remnants: These nebulae are the remnants of massive stars that have exploded as supernovae. They consist of the materials ejected during the explosion and often include shock waves and expanding shells of gas.

Star Clusters

Star clusters are groups of stars that are held together by gravity and all came from the same molecular cloud during stellar birth. Astronomers identify two types of star clusters: **open star clusters** and **globular star clusters**. Star clusters are classified by their size, shape, age, number of stars, and location in the galaxy. Many star clusters are visible to the naked eye, and some can be seen with binoculars or small telescopes.

Globular clusters are spherical-shaped collections of stars that include hundreds of thousands of stars within a few hundred light-years. Living in the halo of our galaxy, they circle the galactic center and are all incredibly old—between twelve and thirteen billion years old. An example of a globular cluster is Messier 13, located in the constellation Hercules. Right at the border of the Keystone asterism in Hercules, it is the brightest globular cluster visible in the northern hemisphere. Though it looks like a fuzzy patch to the unaided eye it is best viewed using binoculars or a telescope. This cluster is estimated to be 145 light-years across and contains over 300,000 stars.

Open clusters do not have a definite shape. They are younger, contain second-generation stars, which are often hot and blue, and live within the disk of our galaxy. An example of an open star cluster is the Pleiades, a young open star cluster. It is located approximately 415 light-years away, and it is probably one of the most famous star clusters, and for good reason. This beautiful cluster appears to be a group of about six stars to the naked eye, but when magnified, many more stars are revealed. The name Pleiades comes from the Greek word *plein*, which means "to sail." This cluster was important to ocean-fairing navigators, and it was used to help sailors determine the best times to sail the Mediterranean Sea. Many cultures around the world have stories and legends about the Pleiades star cluster.

Globular Star Cluster

Messier 13—Great Hercules Cluster
Located in Hercules constellation

Open Star Cluster

Messier 45—Pleiades Star Cluster
Located in Taurus constellation

Galaxies

Galaxies are massive collections of stars, gas, dust, and dark matter bound together by gravity. They are the fundamental building components of the universe and come in a variety of shapes and sizes. Galaxies are critical to the development and spread of stars and planets. Galaxies evolve through interactions and mergers with other galaxies, leading to changes in shape, structure, and star formation activity. Scientists study galaxies by observing them through a range of electromagnetic spectra, from radio waves to gamma rays. This helps to better understand their origin, composition, and behavior.

SPIRAL GALAXIES

When most people think of a galaxy, they typically imagine a spiral shape. Spiral galaxies are distinguished by their flat, spinning disk with a central bulge and spiral arms. The disk contains younger stars, gas, and dust, whereas the central bulge is made up primarily of older stars. They frequently feature a brilliant center core and are further classified as barred spirals (with a central bar-shaped structure) or conventional spirals. The Milky Way, our galaxy, is a barred spiral galaxy.

ELLIPTICAL GALAXIES

Elliptical galaxies range from roughly spherical to elongated in shape. Their look is smooth and featureless. They are older and less active in terms of star production. Elliptical galaxies can range in size from dwarf galaxies to massive ellipticals with billions of stars. Some of the biggest galaxies discovered are elliptical.

A famous example of an elliptical galaxy is Messier 87. This is the galaxy in which scientists captured the first image of a black hole. Messier 87 is part of the Virgo Supercluster of galaxies. It contains several trillion stars and has over 15,000 globular clusters surrounding it. It is estimated to be a whopping 54 million light-years away. It is under the naked eye limit, so a telescope must be used to observe it.

IRREGULAR GALAXIES

Irregular galaxies do not have a distinct, regular shape. These types of galaxies are the youngest of all galaxy types. They contain young stars, gas, and dust and have areas of active star formation. The most well-known examples of irregular galaxies include the Large and Small Magellanic Clouds, which are satellite galaxies of the Milky Way.

The Moon—Our Closest Companion

EARTH'S COMPANION

The moon has indeed played a crucial and influential role in the development of life on Earth. Its presence has had a profound impact on our planet and its living organisms. Here are some key ways in which the moon has influenced life on Earth:

- **Tidal Influence**: The moon's gravitational pull causes tides in Earth's oceans. Tidal movements have shaped coastal ecosystems, influenced marine life, and played a role in nutrient distribution and ocean currents. Tidal zones also provide unique habitats for various organisms.

- **Stabilizing Earth's Axis**: The moon's gravitational influence helps stabilize Earth's axial tilt, which results in stable and moderate climate conditions. This stable axial tilt is crucial for maintaining the seasons and creating a relatively stable environment for life to flourish.

- **Light and Darkness**: The changing phases of the moon, from new moon to full moon and back, create cycles of light and darkness. This has influenced the behavior and reproductive patterns of many organisms, including some nocturnal animals that rely on the moon's light for navigation and hunting.

- **Timekeeping and Human Culture**: The moon's regular cycles provided an early form of timekeeping for ancient civilizations. Many cultures developed lunar calendars to mark important events, agricultural activities, and religious festivals.

- **Eclipse Phenomena**: Solar and lunar eclipses, which occur due to the moon's position between the Earth and the sun, have been observed and recorded by various cultures throughout history. Eclipses have inspired awe and many mythological interpretations of these events.

- **Impact Protection**: The moon has acted as a cosmic shield, protecting Earth from some potentially catastrophic asteroid and comet impacts. Its gravitational pull helps capture some of these objects before they can reach Earth.

- **Scientific Exploration**: The moon has been a target for scientific exploration and a platform for human space missions. Studying the moon's geology, surface, and history has provided valuable insights into the early history of the solar system and Earth.

- **Inspiration and Cultural Significance**: The moon has been a source of inspiration for art, literature, poetry, and folklore across different cultures. It holds cultural and symbolic significance in various mythologies and traditions.

THE MOON AND STARGAZING

Planning your stargazing sessions around the moon cycle is crucial for optimizing your experience. The moon's brightness can significantly impact the visibility of stars and other celestial objects, so being aware of the moon cycle is essential for stargazing success.

New-moon nights are ideal for stargazing because the sky is darkest and faint stars and deep-sky objects become more visible. The moon's light does not interfere with the night sky because the lit portion of the moon is facing away from Earth.

During the few days before and after the new moon, the moon appears as a crescent in the sky. The crescent moon emits less light and doesn't overwhelm the night sky, making it another good time for stargazing. These phases also provide some illumination that can be helpful for navigating in the dark.

In addition to the moon phase, other factors like light pollution, atmospheric conditions, and the specific location from which you're observing also play a role in determining the quality of your stargazing experience. Planning and checking moon phase calendars can help you choose the best nights for observing celestial events.

Stargazing Targets

Most of the objects below can be seen with the unaided eye. Stargazers can use binoculars or telescopes to see more definition and shape of celestial objects.

Great Orion Nebula (Messier 42)

A stellar nursery located 1,344 light-years away. Look near the belt stars of Orion.

Pleiades Star Cluster (Messier 45)

Open star cluster located 440 light-years away. Also called the Seven Sisters.

Hyades Star Cluster (Caldwell 41)

Nearest open star cluster in Taurus is located 153 light-years away. Forms a V-shape.

Double Cluster of Perseus

Two open clusters visible with the naked eye. Located in the Perseus constellation.

Lagoon Nebula (Messier 8)

An emission nebula located 5,800 light-years away. It is close to the Teapot asterism in Sagittarius.

Beehive Cluster (Messier 44)

Open star cluster in Cancer. One of the first objects Galileo studied with a telescope.

Trifid Nebula (Messier 20)

An unusual combination of an open cluster of stars and three types of nebulae.

Rosette Nebula (Messier 6)

This star-forming region stretches 100 light-years across. Located between stars Betelgeuse and Procyon.

Carina Nebula (NGS 3372)

Vast nebula containing multiple star clusters. Best viewed from the southern hemisphere.

Large and Small Magellanic Clouds

Small satellite galaxies of the Milky Way. Best viewed from the southern hemisphere.

Omega Centauri (NGC 5139)

Largest known globular cluster in the Milky Way. Located in the constellation Centaurus.

Andromeda Galaxy (Messier 31)

Closest spiral galaxy to our home galaxy. Located in the Andromeda constellation.

Celestial Events

Meteor Showers

Meteors are fascinating celestial events that captivate viewers around the world. Their brief but brilliant displays of light add a touch of magic to the night sky. The journey of a meteoroid begins in space. When a small piece of debris, often from a comet or asteroid, enters Earth's atmosphere, it becomes a meteor. As it streaks through the sky and burns up due to friction with the air, it is referred to as a meteor. If any part of the meteoroid survives its journey through the atmosphere and lands on Earth's surface, it is then called a meteorite.

While most meteor showers are unpredictable, some are more reliable and occur around the same time each year. Meteor showers are named after the constellation from which they appear to radiate. For example, the Perseids meteor shower seems to originate from the constellation Perseus. However, meteors can be seen in any part of the sky, not just from the radiant point.

Meteor Shower Dates	Peak Dates	Constellation Radiant	Amount of Meteors Per Hour	Source of Debris
Quadrantids Dec 28–Jan 12	January 3–4	Boötes	100	**Asteroid 2003 EH1** This asteroid takes over 5 years to orbit the sun.
Lyrids Apr 14–30	April 22–23	Lyra	20	**Comet C/1861 G1 Thatcher** This comet takes 415 years to circle the sun. The records of this meteor shower go back 2,500 years.
Eta Aquarids Apr 19–May 28	May 5–6	Aquarius	50	**Comet Halley** The earth passes through debris from Halley twice each year and is the source of the Orionids shower.
Perseids Jul 17–Aug 24	August 12–13	Perseus	60–100	**Comet Swift-Tuttle** This comet takes 133 years to orbit the sun.
Orionids Oct 2–Nov 7	October 21–22	Orion	20	**Comet Halley** It takes Comet Halley 76 years to make one orbit around the sun.
Geminids Dec 4–20	December 13–14	Gemini	120–150	**Asteroid 3200 Phaethon** This asteroid takes about 1.4 years to orbit the sun.
Ursids Dec 17–26	December 22–23	Ursa Minor	10	**Comet 8P/Tuttle** It takes 13.6 years for this comet to orbit the sun.

Eclipses

Eclipses are fascinating astronomical events that occur when celestial bodies align in such a way that one body casts a shadow on another. There are two primary types of eclipses: **solar eclipses** and **lunar eclipses**.

SOLAR ECLIPSES

A solar eclipse occurs when the moon passes directly between the earth and the sun, temporarily blocking the sun's light. This alignment can only happen during a new moon. Solar eclipses can be classified into three main types:

1.  **Total Solar Eclipse:** The moon completely covers the sun, revealing the sun's corona. Observers within the path of totality experience complete darkness for a brief period.

2.  **Partial Solar Eclipse:** Only a portion of the sun is obscured by the moon, leading to a partial shadow on Earth.

3.  **Annular Solar Eclipse:** The moon is too far from the earth to completely cover the sun, resulting in a ring-like appearance of the sun around the moon.

Solar eclipses are relatively rare occurrences for any specific location on Earth, as the path of totality is usually narrow. Observers are advised to use proper solar-viewing glasses to protect their eyes during the event.

LUNAR ECLIPSES

A lunar eclipse occurs when the earth passes directly between the sun and the moon, causing the earth's shadow to fall on the moon. This alignment can only happen during a full moon. Lunar eclipses can also be categorized into three types:

1.  **Total Lunar Eclipse:** The entire moon passes through the earth's umbra (the darkest part of its shadow), resulting in a dramatic change in color, often giving the moon a reddish hue, commonly referred to as a "Blood Moon."

2.  **Partial Lunar Eclipse:** Only a portion of the moon enters the earth's umbra, causing part of the moon to darken while the rest remains illuminated.

3.  **Penumbral Lunar Eclipse:** The moon passes through the earth's penumbra (the outer part of its shadow), leading to a subtle shading that is often difficult to observe.

Lunar eclipses are more common than solar eclipses and can be viewed from anywhere in the nighttime side of the earth, making them accessible to a wider audience.

Comets

Comets are fascinating celestial objects composed of ice, dust, and rocky materials. They originate from the outer regions of the solar system, primarily from two key areas: the Kuiper Belt and the Oort Cloud. As comets approach the sun, they undergo significant changes due to solar radiation and the solar wind, which cause their ice to vaporize and release gas and dust. This process forms a glowing coma (a cloud of gas and dust surrounding the nucleus) and often produces a spectacular tail that points away from the sun due to solar wind.

Light pollution can significantly hinder your ability to see comets. Seek dark areas away from city lights for the best viewing experience. Comets are best viewed just before dawn or just after dusk, depending on their position in relation to the sun. The sky should be dark enough to see the faint glow of the comet. Seeing a comet is not a common occurrence like a lunar eclipse or a meteor shower. It is important to keep an eye on astronomical calendars or websites that provide information about upcoming comet appearances. Notable comets often have well-publicized sightings.

STARS

Birth, Life, and Death of Stars

Stars, like humans, have a cycle of birth, life, and death. The entire life of a star cannot be observed within a human lifetime. Stars live much longer than humans, and their life cycle is vastly different than human ones; however, they share some similarities. Both humans and stars are born, grow, and die, and the cycle continues to repeat.

How long a star lives depends on its mass. Its lifespan will be relatively short in comparison with the total age of the universe. The larger and more massive a star is, the faster it will burn through its supply of hydrogen, making its lifespan short. Smaller stars tend to live longer and burn through their energy supply of hydrogen much more slowly than larger stars. A star's mass will also determine its outcome. Smaller stars, such as our sun, will expand in size as their fuel runs out, only to fade out in a brilliant display as they expel their outer layers, leaving behind a hot, leftover core. Massive stars die more violently after they run out of hydrogen fuel in their core. Larger stars will grow into supergiants and then burst into supernovas that leave either a black hole or a spinning neutron star.

Star Formation

A star's existence begins amid a chilly, molecular cloud, often known as a stellar nursery or nebula. These clouds can span many light-years across and are primarily composed of hydrogen molecules with trace amounts of helium and heavier elements. Over time, the cloud may become unstable for a variety of reasons, including a shockwave from a nearby supernova or a collision between molecular clouds. The nebula collapses and condenses, causing the cloud to heat up. As temperatures rise, protostars begin to form as the material shrinks inward, causing an increase in pressure in the system. Some stars will produce an accretion disk, resulting in the formation of planets, asteroids, and dust. When the core of a protostar reaches a temperature of 15 million Kelvin, hydrogen atoms fuse into helium atoms in a process known as fusion. Massive amounts of energy are released into space. Solar winds from the newborn star wipe away any residual dust and debris. The energy produced by fusion exerts an outward pressure that is counterbalanced by gravity's inward pull. The star enters the main sequence phase where it will spend most of its existence.

Main Sequence Stage

A star spends much of its life in this stage, where the forces of fusion and gravity are balanced. For 90 percent of its lifetime, a star maintains a balance between the gravitational force pulling material inward and the thermal pressure from fusion pushing outward. Our sun is currently in this stable main sequence phase. Stars in the main sequence phase are classified by their temperature, color, and luminosity.

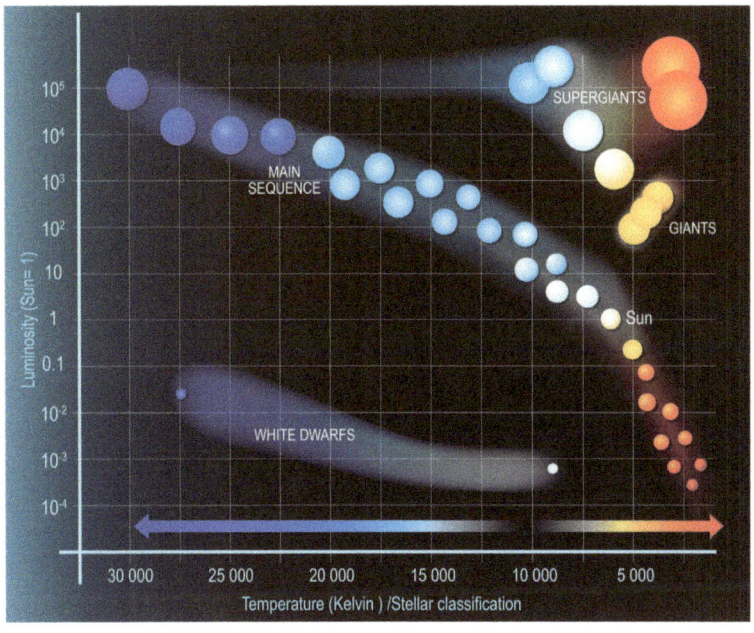

Astronomers use a tool known as the Hertzsprung-Russell diagram (often abbreviated as the HR diagram) to represent stars based on their luminosity (or absolute magnitude) and surface temperature (or spectral type). It depicts the relationship between these traits and helps us to understand the lifespan and classification of stars. The horizontal axis indicates stars' surface temperatures, which decrease from left to right. It can also be classified by spectral type (O, B, A, F, G, K, M). The vertical axis shows the luminosity or absolute magnitude of stars, which increases from bottom to top. The diagonal band extends from the upper left (hot, bright stars) to the lower right (cool, dim stars). Stars positioned above the main sequence have completed hydrogen fusion in their cores, expanded, and cooled to become red giants or supergiants. White dwarfs are in the lower left corner of the picture and are the remains of low- to medium-mass stars that have shed their outer layers and are no longer undergoing fusion. They are extremely hot yet dim due to their small size.

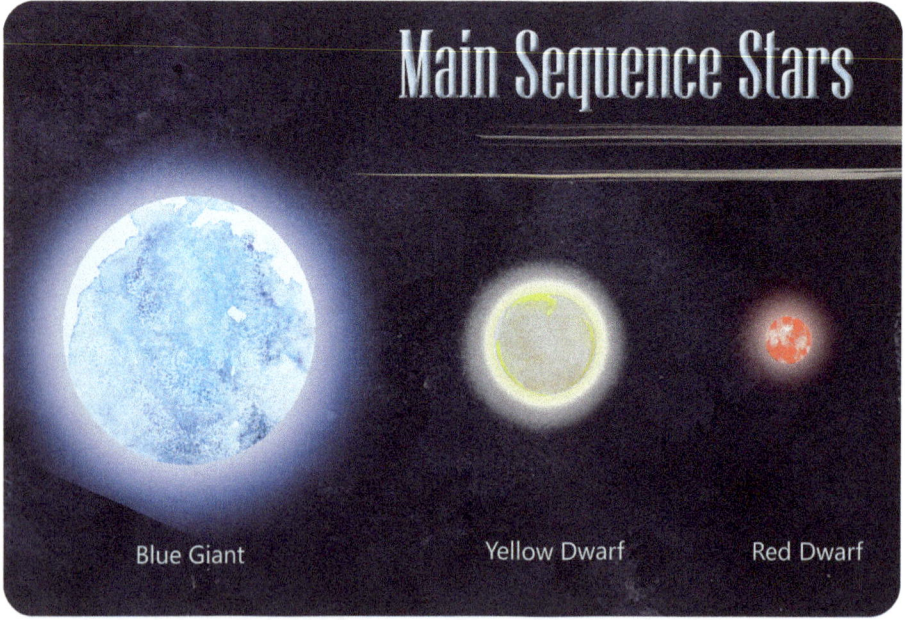

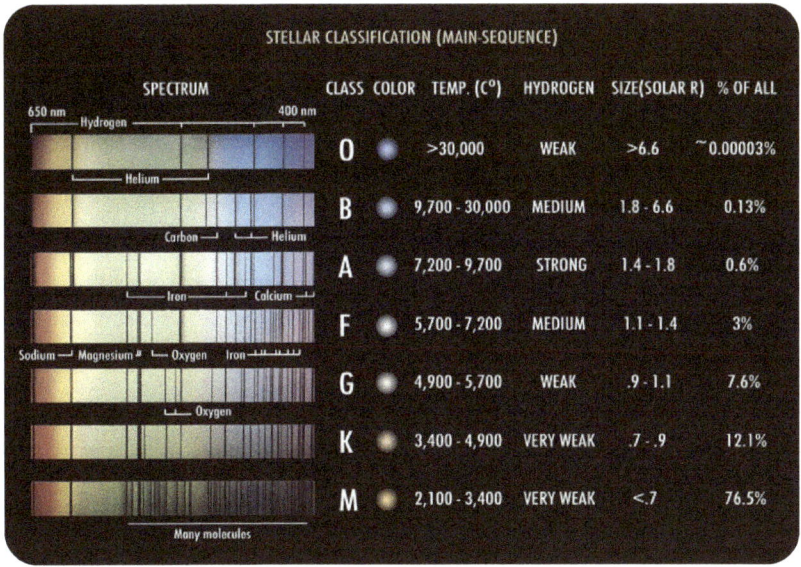

Stellar Death and Remnants

LOW- TO MEDIUM-MASS STARS

Stars that have low to medium mass like the sun will undergo a relatively gentle end to their life cycles compared to more massive stars. When the hydrogen in the core is depleted, fusion stops, and the core contracts while the outer layers expand and cool, forming a red giant. The core temperature rises, allowing helium fusion into carbon and oxygen. The outer layers are eventually ejected, forming a planetary nebula. The gas surrounding the nebula will glow, creating a beautiful, layered structure. The remaining core becomes a white dwarf, which is the dense, leftover core that no longer undergoes fusion. Over time, the white dwarf will radiate away its residual heat and gradually cool, becoming a black dwarf. The cooling process takes billions of years and eventually the leftover core will become cool and invisible.

LARGE-MASS STARS

Massive stars die in a more violent spectacle compared to smaller stars. Their death involves a series of complex and energetic processes that often culminate in a supernova explosion. When the massive star exhausts all the hydrogen in its core, helium fusion begins. In massive stars, fusion does not stop with helium. The core contracts and heats repeatedly, allowing the star to fuse heavier elements in successive stages. First is helium, then carbon, then oxygen, neon, magnesium, and silicon. The final element to form is iron, which cannot be fused into heavier elements. Iron will absorb energy during the fusion process rather than releasing it. As these elements are being fused, the outer layers of the star expand and cool, turning the star into a red supergiant. Some massive stars may become blue supergiants if they retain a high surface temperature.

As iron continues to accumulate in the core, it will eventually reach a critical mass in which the star becomes completely unstable. The core will collapse in a fraction of a second and the outer layers will fall inward. When this happens, a supernova is triggered, causing the star to explode and release shockwaves into the surrounding interstellar space. A supernova is one of the universe's most intense explosions, releasing more energy in a few seconds than the sun would emit throughout its entire ten-billion-year life.

Following the supernova explosion, final stellar death might occur in two ways. Some supernovae will result in a neutron star, which is the remnant core of the star that has condensed into neutrons. Some will collapse into a black hole, which is an object with such a tremendous gravitational attraction that even light cannot escape.

Supernovae events are essential to the formation and dispersion of heavy metals larger than iron in the universe. All the elements on the periodic table that are larger than iron were generated in a past supernova event. The shockwaves that supernovae produce will collide with nearby gas clouds, which will ignite the creation of new stars.

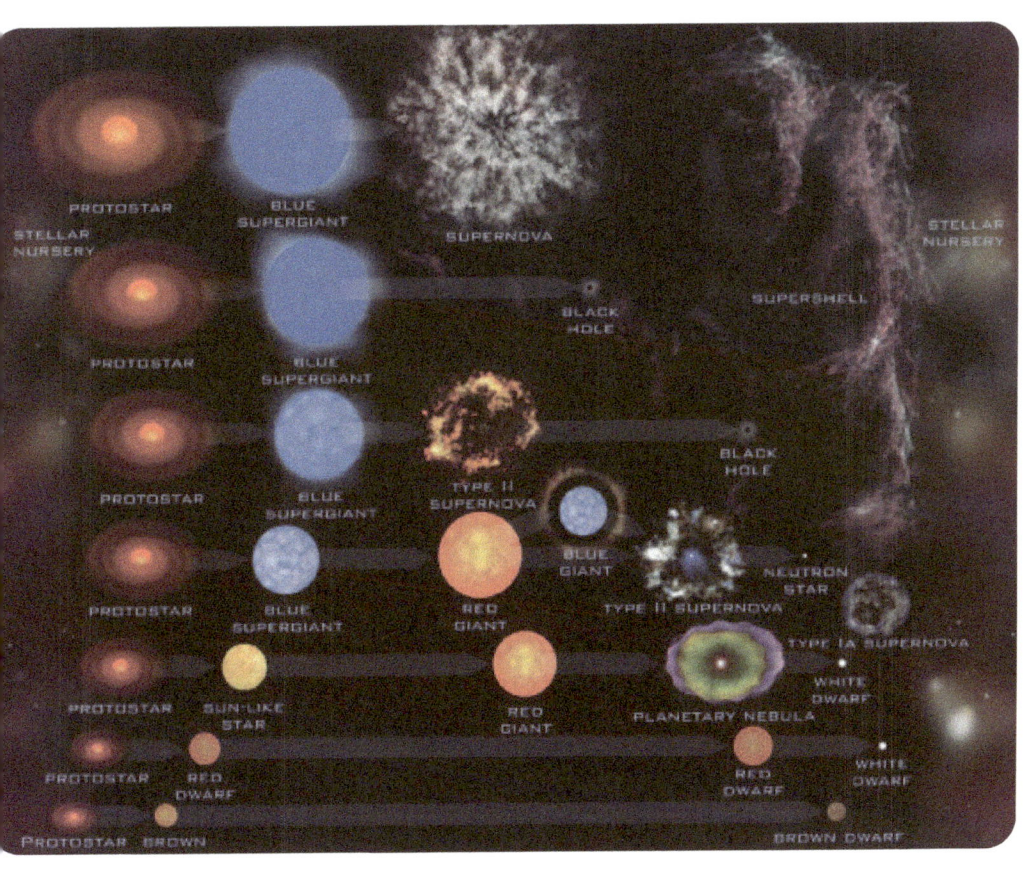

Types of Stars

Protostar: These stars are in the earliest stage of their birth and are hidden inside clouds of gas and dust called nebulae.

T Tauri Stars: T Tauri stars are young, pre-main sequence stars still contracting and approaching the main sequence phase of evolution. They are named after the star in Taurus called T Tauri, the first one to be detected. These stars are commonly found in active star-formation areas, such as molecular clouds and star-forming regions. T Tauri stars often look like main sequence stars in terms of temperature, and the stars are young, less than ten million years. Sometimes these stars can be larger in size than main sequence stars because they haven't fully contracted since fusion hasn't begun. They are variable in their brightness, have strong stellar winds, and can be surrounded by a protoplanetary disk of dust and gas, which are the birthplaces of planets and other celestial bodies.

Main Sequence Stars: A main sequence star is a star in the longest and most stable phase of its life cycle. Hydrogen fuses into helium inside the core of the star. The forces of fusion and gravity are balanced, making the star stable.

Red Dwarf Stars: Red dwarf stars are tiny and relatively cool main sequence stars. They are less massive than our sun and are the most common stars in the universe.

Variable Stars: A variable star is a star whose brightness as seen from Earth (its apparent magnitude) fluctuates over time. These variations can occur over periods ranging from fractions of a second to years or even longer. Certain types of variable stars, like Cepheid variables, have a well-defined relationship between their luminosity and pulsation period. This allows astronomers to determine distances to galaxies and other celestial objects. Studying variable stars provides insights into the life cycles of stars, including stages like the red giant phase and supernova explosions.

Cepheid Variable Stars: Cepheid variable stars, or Cepheids, are pulsing stars that show regular and predictable brightness fluctuations. These pulse periods vary per star. Pulsations can range from days to weeks. Cepheid variables are useful to astronomers because they can act as a distance indicator.

Red Giant Stars: Red giant stars are a late phase in the life cycle of a star that occurs after it approaches the end stages of its life cycle. This stage begins when a stage has exhausted the hydrogen fuel in its core. Fusion begins in the outer layers of the star, causing it to expand. They are larger, cooler, and brighter than most main sequence stars.

Supergiant Stars: Supergiant stars are among the largest and most luminous stars in the universe, representing a late stage in the evolution of massive stars. Supergiants are brilliant, frequently thousands to hundreds of thousands of times brighter than the sun.

White Dwarf: White dwarf stars are the remnants of medium and low-mass stars (those with initial masses less than about four times that of the sun) after they have exhausted their nuclear fuel and shed their outer layers. They are often hotter in temperature but dimmer due to their small size in comparison to other stars.

Black Holes: Black holes are regions in space where the gravitational pull is so strong that nothing, not even light, can escape from them. They are leftover remnants from a large star that exploded into a supernova. Black holes can range in size and mass.

Neutron Stars: A neutron star is a type of compact, extremely dense stellar remnant that forms from the collapsed core of a massive star after a supernova explosion. Neutron stars are extremely dense. A typical neutron star has a mass around 1.4 times that of the sun but a radius of only ten to twelve kilometers (six to seven miles). This makes them so dense that a sugar cube-sized piece of neutron star material would weigh around a billion tons on Earth.

Pulsars: A pulsar is a type of neutron star that rotates and emits beams of electromagnetic radiation from its magnetic poles. As the star rotates, these beams sweep across space. If the beam of radiation is aligned with Earth, they can be detected as regular pulses of radiation.

Magnetars: A magnetar is a type of neutron star with an exceptionally strong magnetic field. Magnetars can produce powerful bursts of X-rays and gamma rays.

Origin of Star Names

The names of stars have changed over the course of human history. Each culture had its own names based upon their own language and histories. Many stars were named after ancient mythological figures or legends. These stories were passed down from generation to generation, preserving cultural wisdom and knowledge of the age. And while the names of the stars may have changed, their patterns of movement have remained the same.

WELL-KNOWN ANCIENT STAR CATALOGS

Astronomer/Source	Civilization/Country	Approximate Date
Babylonian Star Catalog	Mesopotamia	Fifth century BCE
Eudoxus	Greek	Fourth century BCE
Eratosthenes	Greek	Third century BCE
Hipparchus	Greek	Second century BCE
Ptolemy	Greek	Second century CE
Dunhuang Star Chart	Chinese	Seventh century CE
Abd Al-Rahman al-Sufi	Persian	Tenth century CE
Bayer Catalogs	German	Seventeenth century CE
International Astronomical Union	Worldwide (85 participating nations)	Twentieth century CE

In modern astronomy, stars have alphanumeric designations. Astronomers use alphanumeric codes to properly identify the stars they study. The brightest stars and ones of historical, cultural, or astrophysical significance are often given a specific name. Many of these star names are already well-known and have been for a long time, but there was no official IAU collection of names for the brightest stars in our sky until the Working Group on Star Names (WGSN) was founded in 2016. The WGSN is diving into global history and culture to generate a list of officially recognized star names that best represent and respect the global diversity of astronomical knowledge among human cultures both past and present.

Prior to the establishment of the WGSN, the IAU had only formally authorized the names of fourteen stars. The recognized star names by the WGSN have increased over time. To date, the IAU WGSN has officially authorized the names of around 449 stars, with more being considered.

COMMON STAR NAME AND MEANINGS

Star Name/Prefix	Meaning	Example
Rigel	Arabic word for *foot*	+ Rigil Kentaurus—represents foot of Centaurus + Rigel in Orion—represents foot of the Hunter
Deneb	Arabic phrase meaning *tail*	+ Deneb in Cygnus—represents tail of the Swan + Denebola in Leo—represents tail of the Lion

The Sun—Our Closest Star

The Source of All Life

The sun's impact on all life on Earth cannot be understated. All life depends on our home star, providing energy in the form of light and other forms of radiation. The sun is the primary source of energy for all life on Earth. Plants, microbes, and some protists turn sunlight into chemical energy, which is transferred down the food chain through the process of photosynthesis. Solar energy is the backbone of almost all ecosystems. It is easy to see why so many ancient civilizations worshiped the sun.

The sun has a significant impact on Earth's climate and weather patterns. The sun's rays cause unequal heating of the Earth's surface, which causes air circulation and the creation of winds, ocean currents, and weather systems. The sun's energy is responsible for keeping the Earth's average temperature stable and for generating the conditions required for life. Knowing about our own star can help scientists study the changes that may occur with Earth.

The sun is the star that scientists know the most about because of its proximity to Earth. Its patterns, fluctuations, and structure are well documented. As technology advances, scientists will continue to learn about our closest star.

Characteristics of the Sun

The sun is a giant ball of ionized gas called plasma. It is dynamic and ever changing.

The sun does not have a solid surface. It is made completely of ionized gas and it rotates unevenly. The equator of the sun rotates once every twenty-five Earth days while the poles rotate every thirty-six days. The sun, like all the other stars in the galaxy, revolves around the core of the Milky Way. It makes one revolution every 230 million Earth years.

The sun goes through an eleven-year cycle called the solar cycle. Every eleven years, the sun's north and south magnetic poles switch, and this causes high activity in the form of solar storms. Scientists have been studying the sun's patterns since the eighteenth century.

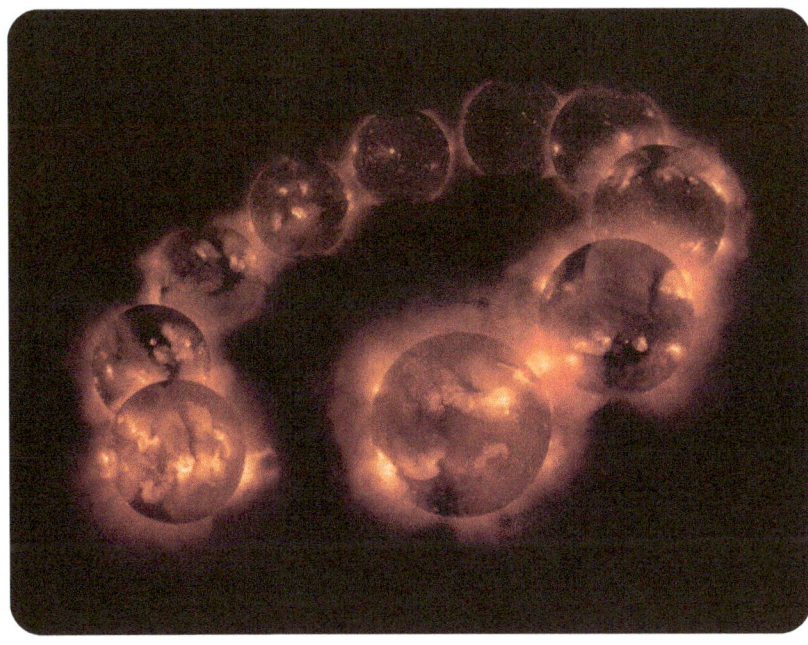

STRUCTURE OF THE SUN

Layer	Characteristics
Core	Innermost region of the sun. Fusion takes place here, which is the nuclear energy process that transforms hydrogen into helium. This process releases vast amounts of energy in the form of heat and light. This process provides fuel for the star. It is the hottest part of the sun at 15,000,000 Kelvin.
Radiative Zone	Layer that extends outwards from the core. Energy produced in the core travels outward via radiation in this zone.
Convection Zone	Extends from the radiative zone to the sun's surface. Energy is transported by convection. Hot plasma rises toward the surface, cools, and then sinks back down to be reheated, creating convection currents.
Photosphere	The visible surface of the sun. The layer from which the sun's light is emitted. It appears granulated due to the convective motions beneath it. This is the only layer that can be observed directly.
Chromosphere	Above the photosphere, it extends up to about two thousand kilometers. This part of the sun can only be seen during a solar eclipse. From this point on, temperatures increase as you move away from the sun.
Corona	Outermost layer, extending millions of kilometers into space. The corona is much hotter than the underlying layers and is the source of the solar wind. It is visible during a total solar eclipse as a white halo.
	One of the great mysteries of the sun is that the temperature of the corona increases a great deal moving away from the chromosphere. Scientists are currently studying this phenomena through the National Aeronautics and Space Administration's Parker Solar Probe.

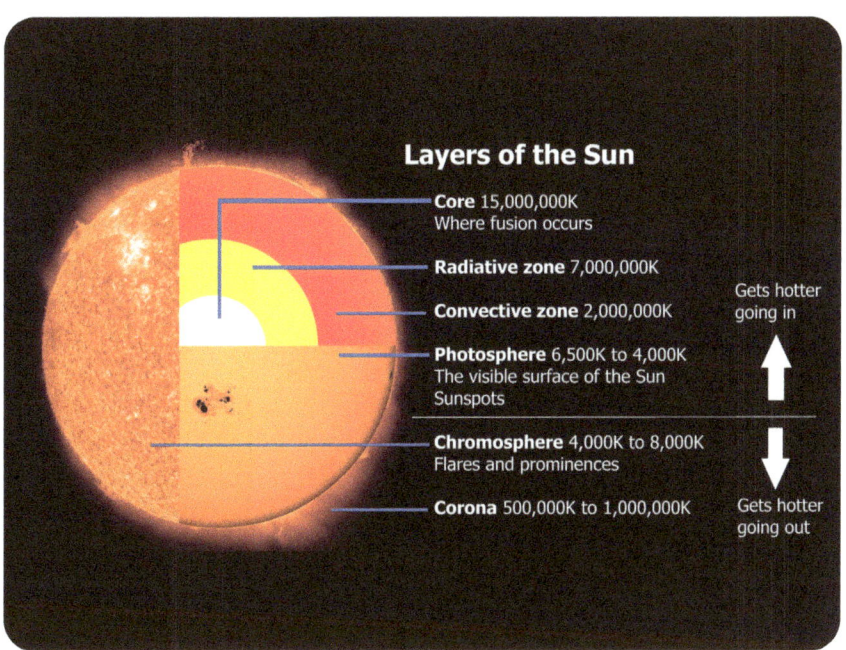

FEATURES OF THE SUN

Sunspots: Darker spots that appear on the surface of the sun. These are cooler regions of the sun that appear to be holes on the surface. The cooler temperatures make them appear darker. These areas are not truly "holes." Sunspots can last anywhere from days to months. These cooler areas on the surface of the sun are due to magnetic disturbances occurring on the surface of the sun.

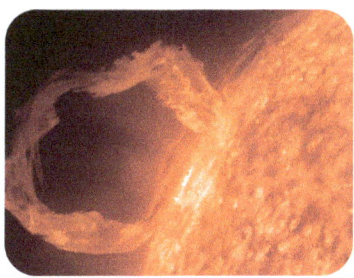

Solar Prominence: A solar prominence is a huge, luminous feature that extends outward from the sun's surface, often in a loop form. These structures are anchored to the sun's photosphere and extend into the corona.

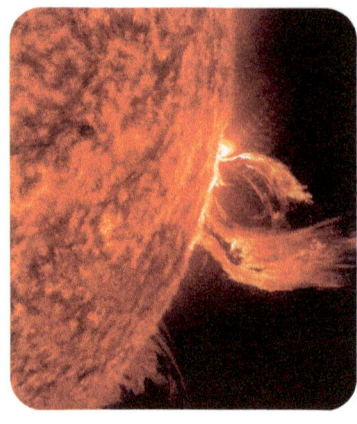

Coronal Mass Ejections: Coronal mass ejections (CMEs) are large bursts of plasma and magnetic fields from the sun's corona, which is the outermost section of the solar atmosphere. They contain particles of subatomic particles that are released from the corona. The frequency of CMEs will vary depending on the solar cycle. There are more common during times of high solar activity. The effects of CMEs on Earth include auroras, geomagnetic storms, satellite and GPS disruptions.

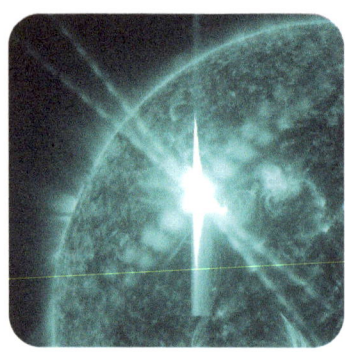

Solar Flares: Solar flares are abrupt and strong bursts of electromagnetic radiation that come from the sun's atmosphere. They are the result of the emission of magnetic energy from the sun's atmosphere. They are similar to CMEs, but are less intense. Solar flares, just like CMEs, can vary in frequency in accordance with the solar cycle. During peak times of the eleven-year solar cycle more solar flares often occur. Sometimes flares can happen in conjunction with CMEs. Solar flares can influence satellite communication and enhance aurora events.

A solar flare can be seen as the bright spot on the surface of the sun. Scientists will study the diffraction patterns of light to learn more about this phenomenon.

Coronal Hole: A coronal hole is a portion of the sun's corona that appears darker, less thick, and cooler than the surrounding sections. Coronal holes are more prominent during periods of low solar activity.

The darker area in the north pole of the sun is a coronal hole.

Solar Wind: The solar wind is a stream of charged particles (plasma) that the sun's corona, or upper atmosphere, releases into space. It is mostly made up of alpha particles, protons, and electrons. A constant stream of plasma that escapes the gravitational attraction of the sun and travels throughout the solar system is known as the solar wind.

The path of the solar wind is imagined in pink. The solar wind is affected by the sun's magnetic field. The form of the spiral that goes up and down has been compared to a ballerina's skirt.

How Do Scientists Study the Sun?

Even though we can't look inside the sun directly, scientists have been able to figure out what's going on inside by using advanced techniques and measurements.

HELIOSEISMOLOGY

Helioseismology is a branch of science that studies the structure of the sun. Scientists study the seismic properties of the star, such as the frequency, wavelength, and speed of these waves traveling across the sun's surface. Oscillations of plasma are caused by sound waves that travel through the sun's interior. By observing these events, scientists can infer details about the sun's internal structure. Helioseismology provides information about temperature, composition, and motion inside the sun. It has confirmed the existence of the different layers and has given precise measurements of their properties.

SPECTROSCOPY

Spectroscopy is the analysis of the light spectra emitted by the sun. By studying the absorption and emission lines in the solar spectrum, scientists can deduce the chemical composition of the sun's outer layers and infer properties of its internal structure. Spectroscopy helps determine the abundances of different elements in the sun and provides clues about the processes occurring beneath the photosphere.

The study of the sun's light spectra is called spectroscopy. Scientists can figure out what chemicals are in the sun's outer layers and what its core structure is like by looking at the absorption and emission lines in its spectrum. Spectroscopy helps figure out how common different elements are in the sun and gives us clues about what's going on below the photosphere.

SOLAR MODELING

Solar modeling is when scientists create detailed computer models based on the laws of physics. These models use input data from observations and fundamental principles of physics to simulate the behavior of the sun's interior. Models can predict how temperature, pressure, and density vary with the depth of each layer. They also help in understanding how energy is transported from the core to the surface through radiation and convection.

GEOMAGNETIC STORMS

Geomagnetic storms are disturbances in Earth's magnetosphere caused by solar wind and solar activity. The interaction between the sun's solar wind and Earth's magnetic field creates some of the most stunning displays of light seen at nighttime.

Events such as solar flares and coronal mass ejections send energized solar wind hurling across space. If these highly charged particles reach Earth's magnetic field, they interact with the magnetosphere of Earth. Effects of these interactions include satellite and communication disruptions, power grid failures, and radiation hazards.

Organizations like NOAA's Space Weather Prediction Center (SWPC) and National Aeronautics and Space Administration (NASA) monitor solar activity and provide forecasts for geomagnetic storms. They use solar observatories in space, satellites to measure solar wind, and ground-based magnetometers to track Earth's magnetic field.

The largest solar event ever recorded was a solar storm known as the Carrington Event, which happened in 1859. It is still the strongest geomagnetic storm ever recorded, and it bears the name Richard Carrington in honor of the British scientist who saw and documented the solar flare connected to the event. On September 1, 1859, Richard Carrington and fellow astronomer Richard Hodgson separately noted a white-light solar flare. Both witnessed and recorded a coronal mass ejection (CME), a great surge of solar energy. In an abnormally short time—about 17.6 hours—the CME intersected Earth. The resultant geomagnetic storm caused significant disturbances in telegraph systems which was the main technology of the time. Auroras could be seen worldwide, including in areas as far south as the Caribbean and Hawaii.

25 BRIGHTEST STARS IN THE NIGHT SKY

Sirius (Canis Major)—The Brightest Star in the Night Sky

Sirius, the brightest star in the night sky, is one that can be seen in almost all parts of the world. It is significant in many cultures in the world, both the past and present. This star is different from our own solar system because it is a binary star system. It contains a white main sequence star (Sirius A) along with a smaller white dwarf star (Sirius B).

Sirius A, the main star of this system, contains twice the mass of the sun and is twenty-five times brighter. Sirius B is the smaller companion star that is at the end of its life. As a white dwarf star, Sirius B is smaller than the size of Earth, yet it is extremely hot. This star is the leftover core of a main sequence star. Sirius A will have the same eventual fate of Sirius B.

The name Sirius comes from the Greek word *seiros*, meaning "scorcher." This term may have derived from Egyptian history, as this star played an important part in the ancient lives of the Egyptians. Whenever Sirius would rise at daybreak, it signaled the time of year when the Nile will flood. This flooding would fertilize the fields, and thus began the Egyptian new year. On the flip side, Sirius would be in the daytime sky during the hot summer months, and the ancient Egyptians believed that heat from Sirius would add to the heat from the sun, thus being dubbed "the scorcher." This might be where the term "the dog days of summer" derived from.

When looking at ancient Indian myths, the stars in Canis Major were believed to form a deer hunter, and the stars in Orion were a stag. Orion's belt was thought to be an arrow sticking in the side of the deer. Sirius was named Tishtrya, a female deity, who was the bringer of rain and water on Earth.

Polynesian culture has a unique story about Sirius. They did not consider it to be the brightest star in the sky. That honor went to another star in the area of constellation Taurus. It is said that Sirius was jealous of this bright star, so he convinced the god Tane to hurl another star at it, which in turn shattered this brilliant star into what is now known at the Pleiades. Aldebaran, which is the bright, red colored star in the eye of Taurus the Bull is the remnant of the star that was hurled across by the deity Tane.

One of the greatest mysteries of Sirius is one that still baffles astronomers today. All of the ancient sources from around two thousand years ago describe Sirius as having a red, or coppery color. But when you observe this star today, it has a whitish-blue color. But curiously enough, around 1000 AD, the story changes. Ancient texts from around the time of 1000 AD reveal the color of the star as being very white with a slight hint of blue. One possible explanation for this change in color was that Sirius's companion star may have been a red giant around two thousand years ago. However, most astronomers agree that the change from red giant to white dwarf would not happen in two thousand years. It would take much longer than in terms of stellar evolution. Plus, gasses that were expelled from the red giant should be able to be detected, but nothing of the sort has been found so far.

STAR CHARACTERISTICS

Star Name	Sirius
Rank	Brightest star in the night sky
Pronunciation	SEER-ee-us
Star Name Meaning	Greek word Seirios, meaning "glowing" or "scorching"
Alternative Names	Dog Star, Canicula, α Canis Majoris
Constellation	Canis Major the Great Dog

Location (Right Ascension, Declination)	06^h 45^m 08^s −16° 42′ 58″
Distance	8.6 light-years
Magnitude	Apparent: -1.46 — Absolute: +1.43
System	Binary star system: Sirius A—white main sequence star Sirius B—white dwarf star
Temperature*	9,900 Kelvin
Spectral Class*	A-type star
Visible From	Northern Hemisphere: 0° to +60°N — Southern Hemisphere: 0° to -90°S
Compared to Our Sun*	Mass: 2 Solar Masses — Luminosity: 25 times brighter than the sun
	Radius: 1.71 Solar Radii
Exoplanets	None detected
Eventual Fate*	Giant star, planetary nebula, then white dwarf
Interesting Facts	✦ Sirius is the brightest star in the night sky. It is two times brighter than the 2nd brightest star, Canopus. ✦ Sirius is so bright in the night sky because it is so close at 8.6 light-years away. ✦ Sirius B, the fainter companion of Sirius A, is smaller than Earth in size. ✦ Sirius B cannot be seen with the unaided eye. Since Sirius A is bigger, it outshines its companion.

*Primary star Sirius A

Canopus (Carina)—Old Man of the South Pole

Canopus is located in the southern constellation Carina, which is often depicted as the keel of the mythological ship Argo Navis from Greek mythology. Canopus is a stunning star known for its brilliance and its role in navigation in the southern hemisphere. Its brightness and distinctive color make it a favorite target for stargazers and astronomers in the southern parts of the world.

Canopus is classified as an A-type supergiant star. It is known for its white or bluish white color which is characteristic of A-type stars. Its apparent magnitude is typically around -0.72, making it the second-brightest star after Sirius. It is especially prominent in the southern hemisphere.

Canopus is located at a distance of approximately 310 light-years from Earth. It is one of the more distant stars visible to the naked eye. Most of the stars that are bright in our sky are close in proximity. Canopus is a massive and luminous star and is several times larger and more massive than our sun.

Canopus has been referenced in various cultures and has had different names throughout history. Its name, "Canopus," is believed to have been derived from the city of Canopus in ancient Egypt. In Egyptian mythology, this store was associated with Osiris. Canopus has been historically used for celestial navigation, especially by ancient seafarers and navigators in the southern hemisphere. For sailors, it served as a reference point for determining latitude.

STAR CHARACTERISTICS

Star Name	Canopus
Rank	2nd brightest star in the night sky
Pronunciation	can-OH-pus

Star Name Meaning	Possibly derived from the city of Canopus in ancient Egypt
Alternative Names	Alpha Carina
Constellation	Carina the Keel
Location (Right Ascension, Declination)	$06^h\ 23^m\ 57^s$ $-52°\ 41'\ 44''$
Distance	Approx. 310 light-years
Magnitude	Apparent: -0.74 — Absolute: -5.71
System	Single star system: Canopus—white supergiant star
Temperature	7,400 Kelvin
Spectral Class	A-type star
Visible From	Northern Hemisphere: 0° to +20°N — Southern Hemisphere: 0° to -90°S
Compared to Our Sun	Mass: 8-9 Solar Masses — Luminosity: 10,000 times brighter than the sun Radius: 71 Solar Radii
Exoplanets	None detected
Eventual Fate	Planetary nebula, then white dwarf
Interesting Facts	✦ Canopus was not visible to the ancient Greeks and Romans, but was visible to the ancient Egyptians and Mesopotamians. This is because of the shift in Earth's axis over thousands of years, which determines which stars can be seen. ✦ This star is featured in the science fiction epic Dune where the desert planet of Arrakis orbits Canopus.

Rigil Kentaurus (Centaurus)— Our Closest Neighbor

Rigil Kentaurus is a triple star system, consisting of three stars: Alpha Centauri A, Alpha Centauri B, and Proxima Centauri. Rigil Kentaurus, often called Alpha Centauri, is a prominent and well-known star system. Rigil Kentaurus is located in the constellation Centaurus located at a distance of about 4.37 light-years from Earth, making it the closest star system to our solar system.

Alpha Centauri A and Alpha Centauri B are two sun-like stars, with Alpha Centauri A being slightly larger and brighter than our sun, while Alpha Centauri B is similar in size to the sun. These two stars are in a close binary system orbiting each other. Proxima Centauri is a red dwarf star and is the closest known individual star to the sun. It is part of the Alpha Centauri system, but it is located at a greater distance from the two sun-like stars.

STAR CHARACTERISTICS

Star Name	Rigil Kentaurus
Rank	3rd brightest star in the night sky
Pronunciation	RYE-jel ken-TAW-russ
Star Name Meaning	Latin for "foot of the centaur"
Alternative Names	Alpha Centauri
Constellation	Centaurus the Centaur
Location (Right Ascension, Declination)	$14^h\ 39^m\ 37^s$ $-60°\ 50'\ 02''$
Distance	Approx. 4.3 light-years
Magnitude	Apparent: +1.33 — Absolute: +4.38

System	Triple star system:	
	Alpha Centauri A—yellow main sequence star	
	Alpha Centauri B—yellow main sequence star	
	Proxima Centauri—main sequence red dwarf star	
Temperature*	5,700 Kelvin	
Spectral Class*	G-type star	
Visible From	Northern Hemisphere: 0° to +25°N	Southern Hemisphere: 0° to -90°S
Compared to Our Sun*	Mass: 1.07 Solar Masses	Luminosity: 1.5 times brighter than the sun
	Radius: 1.12 Solar Radii	
Exoplanets	2 known exoplanets around Proxima Centauri: ✦ Proxima Centauri b ✦ Proxima Centauri c	
Eventual Fate*	Planetary nebula, then white dwarf	
Interesting Facts	✦ Proxima Centauri is technically the closest star to the sun. Its distance is 4.24 light-years away, while the biggest star of the Alpha Centauri system is 4.35 light-years away. ✦ Proxima Centauri cannot be seen with the unaided eye. Its magnitude is 11, which means you would need a telescope to see it. ✦ Ancient cultures would not have been aware of Proxima Centauri because it is too dim for any human to see. ✦ Alpha Centauri A and Alpha Centauri B are the main binary pair in this system. With the naked eye, their combined light looks like one star. Observers can see both stars when looking at the star with a pair of binoculars or a small telescope.	

*Primary star Alpha Centauri A

Arcturus (Boötes)—Guardian of the Bear

Arcturus is one of the top five brightest stars in the sky, and it is located in the constellation called Boötes, the Herdsman. At one point in time, the entire constellation was called Arcturus, but in modern times is now known as the brightest star within the herdsman pattern. The name of this star is derived from the Greek word *arktouros*, meaning "guardian of the bear." This star has often been associated with Ursa Major, also known as the Great Bear constellation. The easiest way to find Arcturus is using the handle of the Big Dipper (asterism inside the larger Ursa Major constellation). You use the arc of the handle to arc over the Arcturus. Memorize this phrase: *Arc to Arcturus*, and you will without a doubt find the famous star of Arcturus.

This star is relatively close to Earth, being only 36.7 light-years away, and is classified as an orange giant star. Scientists speculate that Arcturus has exhausted the hydrogen inside its core, and is now burning the hydrogen in its outer layers, causing the star to expand in size.

The ancient Polynesians identified this star as Hokule'a, or the "Star of Joy." This was the star that was used by the Polynesians to navigate their way across the Pacific Ocean to the shores of the Hawaiian Islands. Launching their double-hulled canoes from Tahiti and Marquesas Islands, these ancient adventures traveled north and east, when they eventually crossed the equator. They continued on until Hokule'a was directly overhead in the summer, and it was from this they were able to determine their latitude. As the voyagers began to travel west due to the trade winds, they eventually landed on the southeastern shores of the Big Island of Hawaii. Since 1976, the Polynesian Voyaging Society's double-hulled canoe, appropriately named after Hokule'a, has completed many journeys using this way-finding style of navigation.

In the Hawaiian Islands, Hokule'a isa zenith star, which is a star that passes directly overhead of the observer during the year.

STAR CHARACTERISTICS

Star Name	Arcturus
Rank	4th brightest star in the night sky
Pronunciation	arc-TOUR-russ
Star Name Meaning	Latinized version of Greek name Arktouros, meaning "guardian of the bear"
Alternative Names	Alpha Boötis, Hōkūleʻa in Hawaiian
Constellation	Boötes the Herdsman
Location (Right Ascension, Declination)	$14^h 15^m 39^s$ +19° 10' 56"
Distance	36.7 light-years
Magnitude	Apparent: -0.05 — Absolute: -0.02
System	Single star system: Arcturus—orange giant star
Temperature	4,280 Kelvin
Spectral Class	K-type star
Visible From	Northern Hemisphere: 0° to +90°N — Southern Hemisphere: 0° to -50°S
Compared to Our Sun	Mass: 1.08 Solar Masses — Luminosity: 170 times brighter than the sun
	Radius: 25 Solar Radii
Exoplanets	None detected
Eventual Fate	Planetary nebula, then white dwarf

Interesting Facts	✦ Arcturus is a star that is moving toward our own star system.
	✦ Arcturus can be found using the Big Dipper Asterism. Use the handle of the dipper to "arc to Arcturus."
	✦ Arcturus is a very important star to the Polynesians and Hawaiian navigators who sailed across the Pacific Ocean. Its name in Hawaiian is called Hōkūleʻa, which means "Star of Joy" and it is a zenith star from Hawaii's latitude.
	✦ Hōkūleʻa, a double-hulled sailing canoe modeled realistically after ancient Polynesian sailing vessels, embarked on a historic voyage, known as "Malama Honua," on May 18, 2014, from Oahu, Hawaii. This voyage was a remarkable three-year circumnavigation of the Earth, aimed at promoting environmental stewardship, cultural exchange, and the revival of traditional Polynesian navigation techniques.
	✦ Arcturus is currently in its red giant phase. It has exhausted its core of hydrogen and is now fusing helium in its core instead. Hydrogen is now fusing in the outer layers of the star, which causes the star to grow in size and cool down.

Vega (Lyra)—Jewel of the Lyre

This bright bluish star in the constellation Lyra is visible in late spring, reaches its peak in the summer, and can be seen far into early autumn. Lyra was represented as a vulture in the oldest tales of Middle Eastern cultures, giving rise to the name Vega for the brightest star, which means "eagle" or "vulture" in Arabic.

Vega is a beautiful star that is close to our own stellar neighborhood. It is a significant star to both present-day astronomers and ancient stargazers. Vega is one of the easiest stars to find in the night sky and can often be identified in areas of heavy light pollution. It was once the pole star In 12,000 BCE, and it will be again in 13,727 due to Earth's precession of the poles.

Vega is located in the constellation of Lyra and is easy to find in the sky. It can be seen in both the northern and southern hemispheres. Vega passes right overhead or near the zenith in the northern hemisphere during the summer months. This makes it a very

straightforward target to find. In the southern hemisphere, you need to look toward the northern horizon, which will be low above that horizon during the winter months. The best way to find Vega is through the Summer Triangle asterism. The Summer Triangle asterism is made up of three stars: Deneb, Altair, and Vega. Vega is the brightest of these three stars.

Vega is the fifth brightest star in the night sky, making it simple to spot, even in locations with high light pollution. For many years, Vega has served as the magnitude standard for astronomers, and it continues to do so. When scientists classify stars, they utilize a color-magnitude scale known as the Hertzsprung-Russell diagram, or HR diagram. It presents a comparison of star color, temperature, and brightness. The Spitzer Space Telescope also determined that Vega has an infrared excess. It is also a variable star, which means its brightness varies with time. Vega is identified as a 10,000-degree Kelvin black body, but its temperature changes. It has a different temperature at its poles than at the equator because it rotates quickly, completing one revolution every 12.5 hours. This quick rotation has an effect on Vega's shape as well. Scientists believe that Vega is flattened at its poles and bulges at its equator. Vega is a standout member of our local group. It is only 26 light-years distant.

Vega has many similarities and differences from our own sun. In terms of mass and radius, Vega is bigger in both size and mass than the sun. It is estimated to be 2.1 times the mass of our sun. Vega is hotter than our sun and 40 times more luminous. Vega is not as spherical as our own sun. It bulges at the equator and is flattened at the poles because of its rapid rotation of 12.5 hours. Both our sun and Vega rotate differently at the equator than at the poles because they are made of gas. They both contain a dust disk. Vega's dust disk is similar to our own Kuiper Belt.

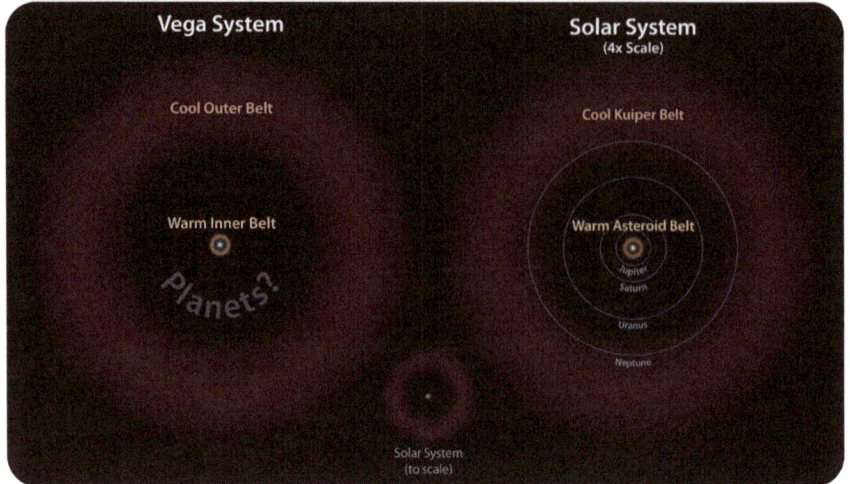

Comparison of the Vega system and our own solar system. Both are surrounded by disks of dust, but the sizes are different.

Importance of Vega to Astronomers

Vega is one of the most studied stars by modern astronomers. Vega was the first star to be photographed (other than our sun) in 1850. Vega was the first spectrographic image that was produced (after our sun) and was one of the first stars in which its distance was measured using the parallax equation. It is also used as a baseline for calibrating the brightness of other stars.

Legends of Vega

The many legends of this star are of both love and tragedy. In Greek mythology, Lyra was depicted as a harp, although some accounts and artwork show a harp wrapped in a vulture's claws. The harp, known back then as a "kithara," was created by the god Hermes. One legend states that Hermes found an empty turtle shell on a beach and strung seven strings through the holes. This instrument created a beautiful sound when played. Hermes later traded his harp to Apollo for a magical staff entwined with snakes, called a caduceus, which had the powers to heal. This staff also gave Hermes the ability to fly.

Another Greek legend tells of how Apollo gave the harp to his son Orpheus, who learned to play the instrument so well that he charmed all the wild beasts he encountered. Later, when his beloved wife Eurydice died, he used his talent with the harp to persuade Hades to release his wife from the underworld. Hades and the other gods agreed, but with the condition that he could not look at her until she reached the upper world. Eurydice was released and Orpheus guided her through the underworld, careful not to look back at her. Just as they were to step into the sun, Orpheus was overcome with joy, and he forgot his promise. He looked back at Eurydice, and she vanished within seconds. Orpheus never recovered from her death and eventually died. Once in the afterworld, Orpheus was reunited with Eurydice, and Zeus placed this magical harp in the sky as a testament to their love. Vega shines as the brightest point in this beautiful collection of stars.

Orpheus guides Eurydice out of the Underworld. There are many variations of this Greek legend.

The Japanese legend of Orihime and Hikoboshi is one of love, loss and reunion. In this legend, the star Vega represents Princess Orihime, and the star Altair represents the cowherd called Hikoboshi. Princess Orihime was a seamstress who wove beautiful clothes by the heavenly river. Orihime worked hard on her weaving, but she became sad and despaired of ever finding love, so her father, god of the heavens, loved her dearly and arranged for her to meet Hikoboshi, the cowherd.
He lived on the other side of the Milky Way. The two fell in love and married. Their passion and devotion to each other were so deep that Orihime stopped weaving, and Hikoboshi allowed his cows to wander the heavens. Orihime's father became angry and forbade the lovers to be together. They tried to reunite, but the celestial river was difficult to cross. But Orihime's father loved his daughter, so he allowed the two lovers to meet once a year on the seventh day of the seventh month.
In Japan, this folktale is celebrated during the Tanabata Festival, also called the Star Festival, where wishes are written on colored paper and hung on bamboo trees. This story is thought to have originated from an ancient Chinese folktale called "The Cowherd and Weaver Girl" (read about the Chinese version in the section about the star Altair).

STAR CHARACTERISTICS

Star Name	Vega	
Rank	5th brightest star in the night sky	
Pronunciation	VEY-ga or VEE-ga	
Star Name Meaning	Arabic name meaning "eagle" or "vulture"	
Alternative Names	Alpha Lyra	
Constellation	Lyra the Lyre, or Harp	
Location (Right Ascension, Declination)	$18^h\ 36^m\ 56^s$ +38° 47' 01"	
Distance	25 light-years	
Magnitude	Apparent: +0.03	Absolute: +0.58
System	Single star system: Vega—white main sequence star	
Temperature	10,000 Kelvin	
Spectral Class	A-type star	
Visible From	Northern Hemisphere: 0° to +90°N	Southern Hemisphere: 0° to -40°S
Compared to Our Sun	Mass: 2.1 Solar Masses	Luminosity: 40 times brighter than the sun
	Radius: 2.5 Solar Radii	
Exoplanets	None detected	
Eventual Fate	Red giant, planetary nebula, then white dwarf	

Interesting Facts	<ul><li>Vega has served as the Earth's pole star because of the precession, or wobble, of the Earth's axis. The wobble of Earth's axis takes 26,000 years to complete one cycle.</li><li>The next time Vega will be the pole star will be the year 12,000 BCE.</li><li>The Vega system is similar to our own system in the sense that it has a warmer, inner rocky belt, much like our own asteroid belt. It also has an outer, dusty debris disk that is cooler in temperature. This area of the Vega system is comparable to our solar system's Oort Cloud.</li><li>Vega is a rapidly rotating star, making its shape different from that of our sun. Vega is flattened at the poles and bulges at the equator. The sun does not have the same shape as Vega because its rotation is slower.</li></ul>

Capella (Auriga)—The Goat Star

Capella is a prominent and easily recognizable star, making it a popular target for stargazers and a useful reference point in the night sky. It appears as a yellowish-white star and is often described as similar in color to the sun. It is a first-magnitude star, which means it is one of the brightest stars visible from Earth. The name "Capella" is of Latin origin and means "little goat." This star is the brightest one in the constellation Auriga the charioteer. Capella is relatively close to our solar system at a distance of about 42 light-years. This makes it one of the closest bright stars to the sun.

Capella is a quadruple star system organized into two binary pairs. The first pair contains two yellow giant stars called Capella Aa and Capella Ab. Both stars are about 2.5 times the mass of the sun, but one is larger than the other. Capella Aa is estimated to be 10 times larger than the sun, while Capella Ab is 8 times larger than the sun. It takes 104 days for this pair to orbit each other, and they are about as far apart as Venus is the sun (0.74 astronomical units). Both of these stars have exhausted their hydrogen cores and have evolved into giant stars. Both stars are fusing heavier elements in the core, and eventually they will turn into planetary nebulae with a hot, leftover core that will become a white dwarf star. Capella Aa and Capella Ab are spectroscopic binaries, which means the stars are too close together to be distinguished as separate points of light, even with powerful telescopes. These types of systems are studied through spectroscopic analysis where astronomers study the properties of light coming from distant stars.

The secondary binary pair consists of two red dwarf stars located much farther away from the main system. Capella H and Capella L orbit the Capella Aa and Capella Ab and are distanced 10,000 astronomical units away from the main pair. The Capella system contains six other visual stars, but these ones are not associated with the main quadruple star system. Capella Aa, Capella Ab, Capella H and Capella L are traveling together in space.

In Greek mythology, Capella is associated with the goat Amalthea (or Amaltheia), the foster mother of Zeus. Amalthea is often depicted as a divine goat that nourished the infant Zeus with her milk when he was hidden in a cave on Mount Ida (in Crete) to protect him from his father, Cronus. In gratitude for her care, Zeus placed her among the stars. The name Capella itself means "small female goat" in Latin, linking the star to this mythological goat.

STAR CHARACTERISTICS

Star Name	Capella	
Rank	6th brightest star in the night sky	
Pronunciation	kah-PELL-ah	
Star Name Meaning	"little goat" or "small female goat" in Latin	
Alternative Names	The Goat Star, Alpha Aurigae, Shepherd's Star	
Constellation	Auriga the Charioteer	
Location (Right Ascension, Declination)	$05^h\,16^m\,41^s$ $+45°\,59'\,52''$	
Distance	42.8 light-years	
Magnitude	Apparent: +0.08	Absolute: +0.29

25 BRIGHTEST STARS IN THE NIGHT SKY

System	Quadruple star system (arranged in binary pairs): Capella Aa—yellow giant star Capella Ab —yellow giant star Capella H—main sequence red dwarf star Capella L—main sequence red dwarf star
Temperature✶	4,970 Kelvin
Spectral Class✶	G-type star
Visible From	Northern Hemisphere: 0° to +90°N / Southern Hemisphere: 0° to -40°S
Compared to Our Sun✶	Mass: 2.5 Solar Masses / Luminosity: 78 times brighter than the sun Radius: 12 Solar Radii
Exoplanets	None detected
Eventual Fate✶	Planetary nebula, then white dwarf
Interesting Facts	✦ Capella Aa and Ab orbit each other every 104 days and their distance away from each other is comparable to our own sun and the planet Venus. ✦ Hawaiians identified Capella as a part of an asterism called Ke ke i Makali'i, which means "the cane bailer of Makali'i." It was an important navigation star to the Hawaiians. ✦ Capella Aa and Ab cannot be resolved separately with an ordinary telescope. The only telescope that has been able to capture these two stars is the Cambridge Optical Aperture Synthesis Telescope (COAST) located in Cambridge, England. ✦ There are 3 stars that lie very close to Capella called "the Kid" stars and it is a well-known asterism. ✦ The Capella star system is one of the brightest X-ray sources in the sky.

✶Primary star Capella Aa

Rigel (Orion)—Foot of the Giant

Rigel, also known as Beta Orionis, is the brightest star in the constellation of Orion. This star carries historical and cultural significance in various traditions around the world. The name "Rigel" is derived from the Arabic word "Rijl," which means "foot." It is associated with the star's position in the constellation Orion as one of the "feet" or "hooves" of the celestial hunter. Rigel is a multiple star system. It has at least three faint companion stars, but they are challenging to observe due to the brilliance of Rigel itself. Rigel is part of a stellar system that contains at least four components. Rigel A is the largest star and the one that is closest to becoming a supernova.

Rigel A, the largest and brightest star of this system, is an extraordinarily large star, estimated to have 17 times the mass of our sun. This star is incredibly bright, with a luminosity around 120,000 times that of the sun. Its high luminosity is due to its massive size and intense energy output. Rigel's surface temperature reaches up to 10,000 Kelvin (about 18,000 degrees Fahrenheit). This high temperature produces its distinctive blue-white color.

The secondary star, called Rigel B, is a spectroscopic binary star with stars Ba and Bb. Both are blue main sequence stars that are approximately three times more massive than our sun. These two stars cannot be resolved through an optical telescope. They are spectroscopic binary stars, which means they can only be detected by analyzing a star's light. Rigel C orbits the Rigel B binaries, forming a close triple system. Since these three stars are so close together, it is difficult to understand each stars' individual properties. Rigel A is by far the most dominant star of the system in terms of size, however, it is far away from its companions at a distance of 2,200 astronomical units. It is estimated that Rigel A and the triple Rigel BC system have an orbital period of 24,000 years. As time goes on and our technology continues to advance, we will be sure to learn more about Rigel and each individual star in this system.

Rigel A, the largest star of this quadruple system, has reached an advanced stage in its stellar evolution. Massive stars like Rigel A burn through their nuclear fuel more quickly than smaller stars like the sun. This quick consumption of fuel leads to short life in the order of millions of years. Stars like Rigel have already burned through their primary fuel sources, such as hydrogen and helium, and have begun fusing heavier elements. This fusion process generates tremendous energy, which sustains the star's luminosity and prevents it from collapsing under its gravity. When nuclear fusion ceases in the core, gravity causes the core to collapse rapidly and trigger a supernova explosion. The outer layers of the star are violently expelled into space, producing an extremely bright burst of radiation that can outshine entire galaxies. The core of the star may collapse further, forming either a neutron star or, in the case of the most massive stars, a black hole. This will be the eventual fate of Rigel A.

STAR CHARACTERISTICS

Star Name	Rigel	
Rank	7th brightest star in the night sky	
Pronunciation	RYE-jel	
Star Name Meaning	Arabic for "foot"	
Alternative Names	Beta Orionis	
Constellation	Orion the Hunter	
Location (Right Ascension, Declination)	05^h 14^m 32^s -08° 12' 06"	
Distance	860 light-years	
Magnitude	Apparent: +0.18	Absolute: -7.84

System	Quadruple star system:	
	Rigel A—blue supergiant star	
	Rigel Ba—main sequence blue giant star	
	Rigel Bb—main sequence blue giant star	
	Rigel C—main sequence blue giant star	
Temperature*	12,000 Kelvin	
Spectral Class*	B-type star	
Visible From	Northern Hemisphere: 0° to +85°N	Southern Hemisphere: 0° to -75°S
Compared to Our Sun*	Mass: 18 Solar Masses	Luminosity: 120,000 times brighter than the sun
	Radius: 17 Solar Radii	
Exoplanets	None detected	
Eventual Fate*	Supernova, then neutron star or black hole	
Interesting Facts	✦ Rigel is a quadruple star system. The primary star is a blue supergiant, while the other three stars form a triple system of blue main sequence stars. All four stars orbit a common center of gravity. ✦ Rigel is a younger star, estimated to be 7 to 9 million years old, where our own sun is 5 billion years old. ✦ Rigel is a variable star, which means its brightness can change over time.	

*Primary star Rigel A

Procyon (Canis Minor)— The Lesser Dog Star

Procyon is located in the constellation Canis Minor, which is often depicted as a small dog or the "lesser dog" following the larger constellation, Canis Major. It is located at a distance of approximately 11.46 light-years from Earth. It is one of our closest stellar neighbors and is part of the group of stars known as the "Local Bubble." The Local Bubble is a hot, low-density gas area in the Milky Way galaxy's interstellar medium, which contains our solar system. It is estimated to have originated as a result of a sequence of supernova explosions that happened between ten and twenty million years ago, leaving a cavity in the surrounding interstellar material.

Procyon is a binary star system, with the two stars orbiting around a common center of mass. The primary star, Procyon A, is significantly brighter and more massive than its companion, Procyon B. The pair orbits each other every forty years. Procyon A is a late-stage main sequence star and is estimated to evolve into its next stage of life, a giant star. When this happens, Procyon A will grow in size and cool down, which will change its appearance from white hue to a reddish orange one. Procyon B is a white dwarf star that contains half the mass of the sun. This is the final stage for a main sequence star. Once Procyon B cools down, it will no longer give off light and become invisible.

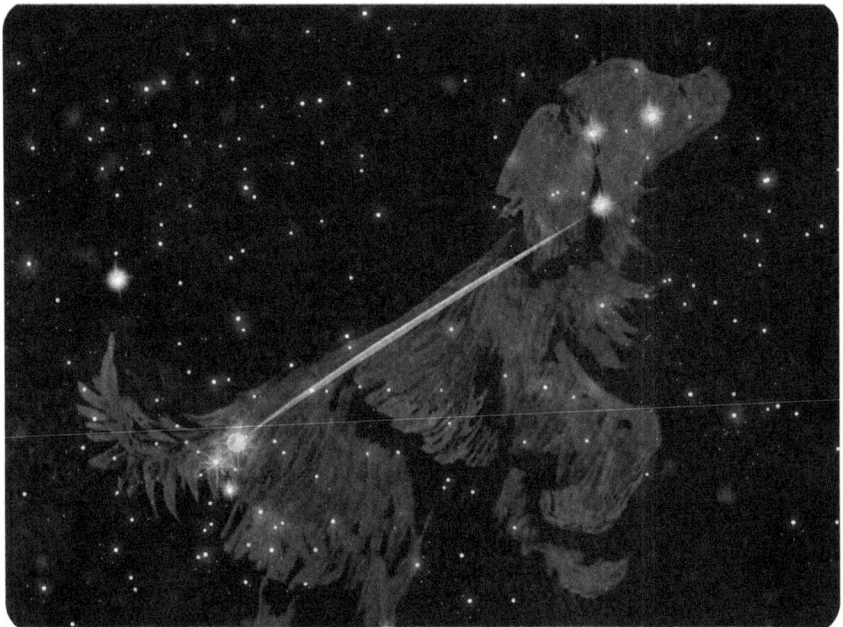

The best time to observe this star is from early December until April. Procyon is located in the constellation Orion and can be found by referencing Orion's belt. Draw a line from Bellatrix to Betelgeuse and continue it to Procyon. Procyon is also a vertex in the Winter Triangle asterism. The last two points of the Winter Triangle are Sirius (the brightest star in the sky) and Betelgeuse.

Use the shoulder stars of Orion (upper right) as pointers to find Procyon (upper left). The brightest star is Sirius (bottom center).

STAR CHARACTERISTICS

Star Name	Procyon
Rank	8th brightest star in the night sky
Pronunciation	PRO-see-on
Star Name Meaning	Greek meaning "before the dog"
Alternative Names	Alpha Canis Minoris, HR 2943, HD 61421
Constellation	Canis Minor the Lesser Dog
Location (Right Ascension, Declination)	$07^h\ 39^m\ 18^s$ 05° 13' 30"
Distance	11.46 light-years

25 BRIGHTEST STARS IN THE NIGHT SKY

Magnitude	Apparent: +0.37	Absolute: +2.66
System	Binary star system: Procyon A—main sequence yellow-white star Procyon B—white dwarf star	
Temperature*	6,500 Kelvin	
Spectral Class*	F-type star	
Visible From	Northern Hemisphere: 0° to +90°N	Southern Hemisphere: 0° to -75°S
Compared to Our Sun*	Mass: 1.5 Solar Masses	Luminosity: Approx. 7 times brighter than the sun
	Radius: 2.0 Solar Radii	
Exoplanets	None detected	
Eventual Fate*	Red giant, planetary nebula, then white dwarf	
Interesting Facts	✦ Procyon got its name "before the dog" because it used to rise much earlier than Sirius in the times of ancient Greece. ✦ Procyon was connected to the mythological figure of Anubis in ancient Egyptian times. ✦ The Objibwe of North America grouped Procyon with Orion to make a larger constellation called the Wintermaker. ✦ Procyon is part of the asterism called the Winter Triangle, which connects this star with Betelgeuse in Orion and Sirius in Canis Major. ✦ The Procyon Star System is similar to that of Sirius—both are binary star systems that have one main sequence star and a white dwarf star.	

*Primary star Procyon A

Achernar (Eridanus)— End of the River

Achernar, often called Alpha Eridani, is the brightest star in the constellation Eridanus the River. Eridanus (pronounced ih-RID-un-us) is one of the largest constellations in the night sky, representing a winding celestial river. Its name is derived from the Latin word for river, and it has roots in Greek mythology. This constellation is a long, twisting line of stars that starts near the bright constellation Orion and winds southward across the sky, symbolizing a flowing river. Achernar is the brightest point found at the end of the winding river.

Achernar is the ninth brightest star in the night sky and is a binary star system consisting of two large blue main sequence stars. This system is estimated to be 139 light-years away. This star system is best viewed from the southern hemisphere, particularly during the month of November, however, it can be seen in the northern hemisphere from between 32 degrees north to the equator.

The main star, Achernar A, is a class B blue star that has begun to evolve off the main sequence band and will eventually expand into a supergiant star. It also has an extremely fast rotational speed causing it to be a flattened sphere. This extreme rotation makes it difficult to measure stellar temperature. It is estimated that the poles are much hotter than the equator. Achernar B is the smaller of the two stars, yet is double the size of the sun. These two stars orbit each other every seven years.

STAR CHARACTERISTICS

Star Name	Achernar
Rank	9th brightest star in the night sky
Pronunciation	AK-er-nar
Star Name Meaning	Arabic for "end of the river"
Alternative Names	Alpha Eridani
Constellation	Eridanus the River
Location (Right Ascension, Declination)	$01^h\ 37^m\ 42^s$ $-57°\ 14'\ 12''$
Distance	139 light-years

Magnitude	Apparent: +0.46	Absolute: -1.46
System	Binary star system: Achernar A—main sequence blue giant star Achernar B—main sequence blue giant star	
Temperature*	12,000 Kelvin	
Spectral Class*	B-type star	
Visible From	Northern Hemisphere: 0° to +32°N	Southern Hemisphere: 0° to -90°S
Compared to Our Sun*	Mass: 6 Solar Masses	Luminosity: Approx. 3,500 times brighter than the sun
	Radius: 6-9 Solar Radii	
Exoplanets	None detected	
Eventual Fate*	Red giant, planetary nebula, then white dwarf	
Interesting Facts	✢ Archenar has an extreme rotational speed, making it a flattened sphere. ✢ The poles of Archenar are hotter than the equator. This is due to its extreme rotation. ✢ A few other stars have a flattened shape like Archenar. Altair and Regulus are also oblate spheroids due to their quick rotation. ✢ Archenar was not visible to the ancient Greeks due to the precession of the Earth. Precession is a slow and continuous change in the orientation of Earth's rotational axis. Over the course of 26,000 years, the orientation of Earth's axis makes a complete circle, pointing to different stars over the course of this timeframe. ✢ The best time to view this star is from the southern hemisphere in November. It can be seen in the northern hemisphere, but only from 32 degrees latitude and lower. ✢ The two stars in this system orbit each other every seven years.	

*Primary star Achernar A

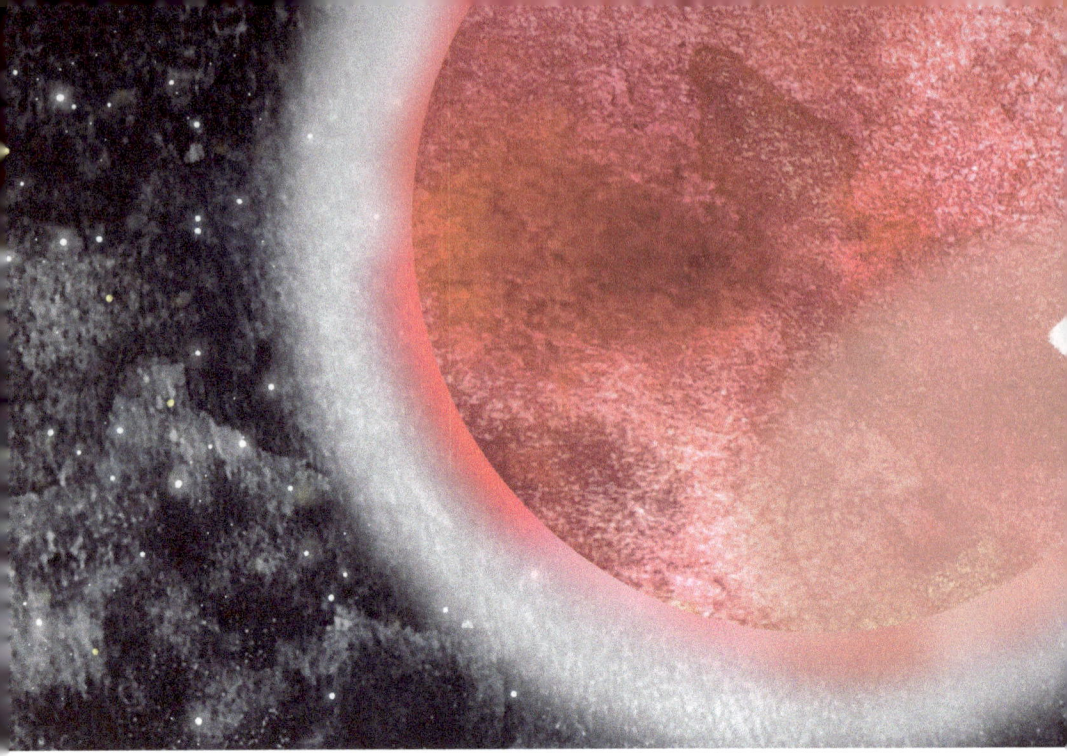

Betelgeuse (Orion)—Red Supergiant in the Hunter

Betelgeuse, also known as Alpha Orionis, is located in the constellation Orion and marks the shoulder of the hunter. Its unique red color and its position in the night sky have made it a well-known and easily recognizable star.

Betelgeuse is a red supergiant star. It is among the largest and most massive stars known. The distance is estimated to be around 640 light-years from Earth. Betelgeuse has a distinct reddish or orange color, which is characteristic of red supergiants. Its color is due to its relatively low surface temperature. Betelgeuse is known to be a variable star, meaning its brightness changes over time. These variations are not entirely predictable, but they can make Betelgeuse appear brighter or fainter in the night sky.

Betelgeuse was once an O-type main sequence star, but it has grown into a supergiant star and consuming nuclear fuel at an astounding rate. It is one of the closest stars to Earth that is in the later stages of its life cycle and it is expected to eventually explode in a supernova, although the timing of this event is uncertain and could be thousands or even millions of years in the future.

In 2019, the brightness of Betelgeuse drastically changed, putting scientists on alert that perhaps it was this star's time to explode. The star grew noticeably dimmer over the weeks. However, the star eventually returned to its normal brightness. Researchers have determined that dimming was caused by the star's atmosphere releasing a huge amount of gas. This material surrounded the star and eventually cooled, creating a cloud that blocked some of the star's light. Astronomers have called this event "The Great Dimming" of Betelguese.

Betelgeuse is easy to find in the night sky as a part of Orion the Hunter. This pattern is one of the most recognizable star patterns in the world and can be seen from both hemispheres. Betelgeuse is the reddish hued star that represents one of the shoulders of the hunter.

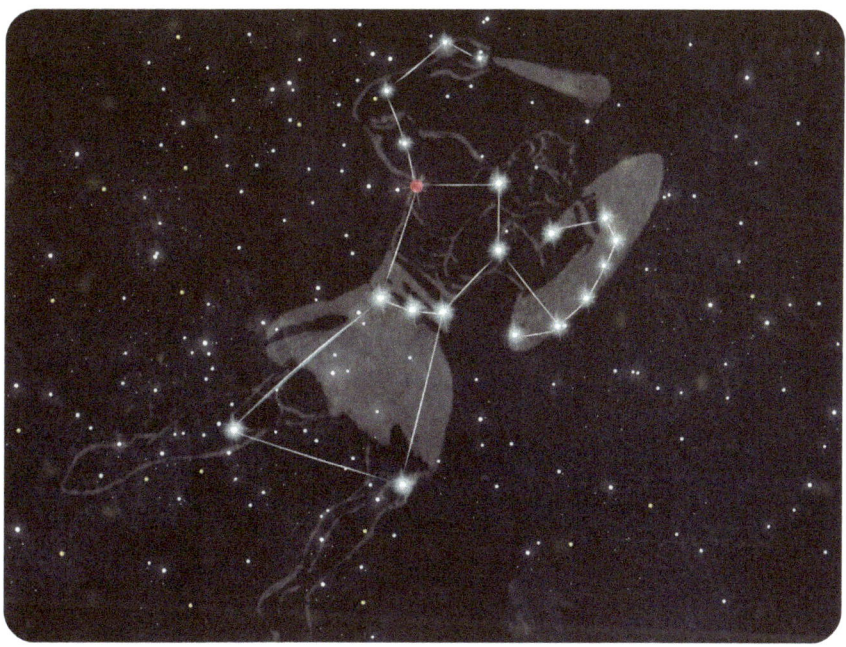

STAR CHARACTERISTICS

Star Name	Betelgeuse	
Rank	10th brightest star in the night sky	
Pronunciation	BET-el-jooz	
Star Name Meaning	Arabic words "Ibt al-Jauzā'" which translates to "the armpit of Orion." This name has changed over time through translations across different languages.	
Alternative Names	Alpha Orionis	
Constellation	Orion the Hunter	
Location (Right Ascension, Declination)	05^h 55^m 10^s +07° 24' 25"	
Distance	640 light-years	
Magnitude	Apparent: +0.4	Absolute: -5.85
System	Single star system: Betelgeuse—red supergiant star	
Temperature	3,600 Kelvin	
Spectral Class	M-type star	
Visible From	Northern Hemisphere: 0° to +85°N	Southern Hemisphere: 0° to -75°S
Compared to Our Sun	Mass: 16–19 Solar Masses	Luminosity: Over 100,000 times brighter than the sun
	Radius: 760 Solar Radii	
Exoplanets	None detected	
Eventual Fate	Supernova, then black hole or neutron star	

Interesting Facts	· Betelgeuse is a variable star, meaning it changes its brightness over time. This makes it difficult to determine its true size and luminosity. · Betelgeuse is a gigantic star when compared to the sun. Its size is bigger than the orbit of Jupiter! · Betelgeuse is a young star. Astronomers estimate its age to be 8 million years old, which seems old to us humans, but relative to the age of the universe (14–15 billion years old), this is a very short time span. · Betelgeuse is one of the few stars in which a direct image of its surface has been obtained. It was first imaged by the Hubble Space Telescope in 1995.

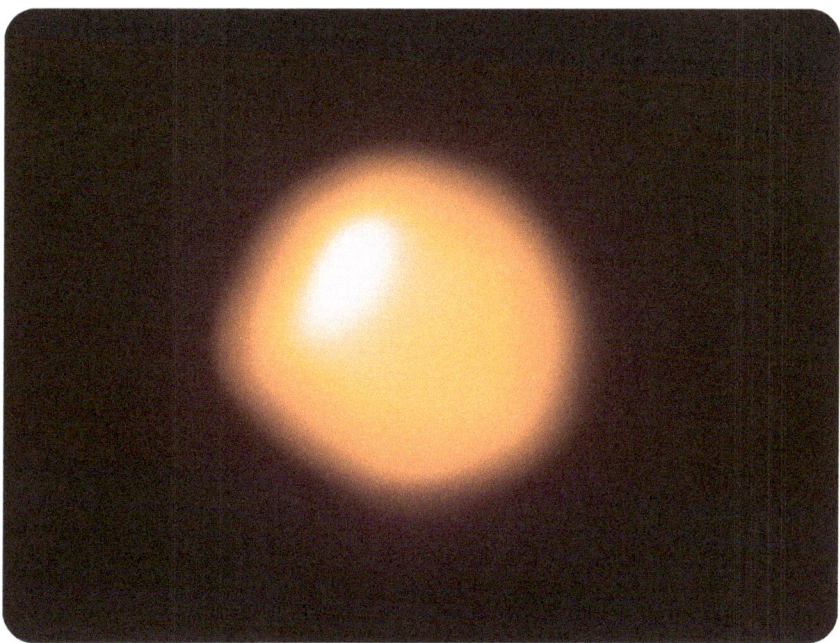

This high-resolution image reveals the surface of Betelgeuse, as seen by the Atacama Large Millimeter/submillimeter Array (ALMA) in Chile.

Hadar (Centaurus)— Triple Star System

Beta Centauri, commonly known by its traditional name Hadar, is a bright star in the southern constellation Centaurus. Hadar is classified as a B-type blue giant star. Hadar, together with Alpha Centauri (Rigil Kentaurus), form the "Pointers" of the Southern Cross constellation.

Although this star looks like one point in the night sky, it is actually a triple star system made up of all B-class blue main sequence stars. Beta Centauri Aa is the largest star of this system and makes a binary pair with Beta Centauri Ab. Both of these stars have begun to evolve out of the main sequence stage and are in the early stages of becoming supergiant stars. As both stars exhaust their fuel, fusion will take place in the outer layers, causing both stars to grow in size and evolve into supergiants. Eventually nuclear fuel will run out causing the stars to explode, leaving behind either neutron stars or black holes.

Beta Centauri Aa and Ab orbit each other every 357 Earth days. The third star, called Beta Centauri B, is also a B-class blue star, but is smaller and further away (210 astronomical units) from the main pair. It takes 600 days for Beta Centauri B to orbit the main pair.

STAR CHARACTERISTICS

Star Name	Hadar
Rank	11th brightest star in the night sky
Pronunciation	HAH-dahr
Star Name Meaning	Arabic meaning "to be present"
Alternative Names	Beta Centauri, Agena
Constellation	Centaurus the Centaur
Location (Right Ascension, Declination)	$14^h\ 03^m\ 49^s$ $-60°\ 22'\ 22''$
Distance	390 light-years

25 BRIGHTEST STARS IN THE NIGHT SKY

Magnitude	Apparent: +0.61	Absolute: -4.9
System	Triple star system: Beta Centauri Aa—main sequence blue giant star Beta Centauri Ab—main sequence blue giant star Beta Centauri B—main sequence blue giant star	
Temperature*	25,000 Kelvin	
Spectral Class*	B-type star	
Visible From	Northern Hemisphere: 0° to +25°N	Southern Hemisphere: 0° to -90°S
Compared to Our Sun*	Mass: 12 Solar Masses	Luminosity: 66,000 times brighter than the sun
	Radius: 9 Solar Radii	
Exoplanets	None detected	
Eventual Fate*	Supergiant, supernova, then black hole or neutron star	
Interesting Facts	✦ Alpha Centauri and Beta Centauri are both pointers to Crux, the Southern Cross constellation. ✦ Alpha Centauri and Beta Centauri are both triple star systems. ✦ The traditional name of Agena possibly comes from the Latin name *genua*, which means "knees." This star represents the knee of the Centaur. ✦ Hadar has played an important role in many cultures in the southern hemisphere. ✦ Stars Aa and Ab are much more massive than Centaurus B. Centaurus Aa is 12 solar masses and Centaurus Ab is 10 Solar masses. These two stars are likely to become future supergiants. Centaurus B clocks in at 4 solar masses. ✦ All three stars in the system have the same spectral class: B-type blue stars.	

*Primary star Beta Centauri Aa

Altair (Aquila)—Eye of the Eagle

Altair is a remarkable star; its shape is unique, it is extremely close to us, and it has been well known by numerous civilizations throughout human history. This star is associated with a variety of ancient mythologies, many of which are still celebrated in cultural festivals today.

Altair is the alpha star of the constellation Aquila, making it the brightest star in the constellation and the 12th brightest star in the night sky. It is estimated to be 16.7 light-years distant, making it a close star in our local stellar neighborhood. Altair is a class A main sequence star, hence it appears to be bluish-white in color. Altair is larger than our sun and has a unique shape because of its very fast rotation. It spins every nine hours, giving it an oblate spheroid form, which is more oblong toward the equator and slightly flattened at the poles. Our sun, in contrast, revolves every twenty-five to thirty-five days, giving it a more spherical form.

Mythologies of Altair

Altair is associated with the Japanese myth of Orihime and Hikoboshi, a famous East Asian story that is particularly prominent in Japanese culture, where it is associated with the Tanabata Festival (see Vega section). This story originated from a Chinese legend called the "Cowherd and Weaver Girl" and tells of the tale of lovers Zhinü and Niulang.

The fable of Zhinü and Niulang is a well-known and popular story in Chinese mythology, frequently connected with the Qixi Festival, also known as the Double Seventh Festival. This romance story has been passed down through centuries. In this love story, the Weaver Girl, Zhinü, is a celestial being and the seventh daughter of the Jade Emperor, who is the ruler of Heaven. She is skilled at weaving beautiful, intricate fabrics. Niulang, the cowherd, is a kind and hardworking mortal man. Zhinü is represented as the star Vega, and Niulang is represented by the star Altair.

The legend goes that Zhinü, tired of her celestial duties, descends to Earth for a break. She meets Niulang, who is living a simple and honest life as a cowherd. They fall deeply in love and get married, living happily together and having two children, a son and a daughter. When the Jade Emperor discovers that his daughter has married a mortal, he is furious. He orders Zhinü to return to Heaven, and when she disobeys, she is forcibly taken, leaving Niulang heartbroken. With the help of a magical ox that Niulang had been caring for, Niulang and his children fly to Heaven to reunite with Zhinü. However, the Queen Mother of the West (another powerful celestial figure) creates a wide river, the Milky Way, to separate them. Touched by their love and devotion, the magpies of the world form a bridge across the river once a year, on the seventh day of the seventh lunar month, allowing Zhinü and Niulang to meet. Deneb is seen as a guiding star to their annual reunion.

The Qixi Festival falls on the seventh day of the seventh month of the Chinese lunar calendar. Young women customarily show their needlework and weaving abilities during Qixi, which are related with Zhinü's qualities. Couples pray for love and happiness in their unions as well. Throughout Chinese history, innumerable poems, paintings, and literary works have been inspired by the tale of Zhinü and Niulang. There are many versions of this love story. In Japan, the lovers are known as Hikoboshi (Altair) and Orihime (Vega) in the ancient legend and the Japanese Tanabata Festival celebrates this event. The Chinese legend predates the Japanese legend, and each has their own versions of the folktale. This enduring love story continues to be celebrated across Asia.

STAR CHARACTERISTICS

Star Name	Altair	
Rank	12th brightest star in the night sky	
Pronunciation	AL-tair	
Star Name Meaning	Arabic name meaning "eagle"	
Alternative Names	Alpha Aquilae, Eagle star, "pillar of heaven" in Māori	
Constellation	Aquila the Eagle	
Location (Right Ascension, Declination)	$19^h\ 50^m\ 47^s$ +08° 52' 06"	
Distance	16.7 light-years	
Magnitude	Apparent: +0.76	Absolute: +2.22
System	Single star system: Altair—white main sequence star	

Temperature	6,800 to 8,600 Kelvin	
Spectral Class	A-type star	
Visible From	Northern Hemisphere: 0° to +90°N	Southern Hemisphere: 0° to -75°S
Compared to Our Sun	Mass: 1.8 Solar Masses	Luminosity: 11 times brighter than the sun
	Radius: 2 Solar Radii	
Exoplanets	None detected	
Eventual Fate	Red giant, planetary nebula, then white dwarf	
Interesting Facts	<ul><li>Altair has an oblong shape because it is rotating so quickly. It spins every nine hours on its axis.</li><li>Altair is one of the few stars that have been directly imaged. The CHARA array of the Mount Wilson observatory in California imaged the surface of Altair in 2006.</li><li>Altair is one of the stars that makes up the asterism called the "Summer Triangle." It marks one of the vertices of this false constellation. Vega and Deneb make up the other two points of the triangle. The Summer Triangle can be used to find other constellations in the sky.</li><li>Since Altair has a rapid rotation, its surface temperature varies. Astronomers have estimated that the equator has a cooler temperature than the poles of the star.</li></ul>	

Acrux (Crux)—Brightest of the Southern Cross

Acrux is the brightest star in the constellation of Crux, the Southern Cross and is the 13th brightest star overall in the night sky. This star system is composed of five stars. The primary components are Acrux A, Acrux B, and Acrux C. Acrux A and Acrux B are closer in distance, while Acrux C is farther away.

Telescopically, this star can be resolved into three components: Acrux A, Acrux B, and Acrux C. But looking at it spectroscopically, more stars appear in this system. Acrux A and Acrux C are spectroscopic binaries, which means these companions cannot be resolved through a telescope, but rather through the analysis of a star's light. When looking at this star through a telescope, only three stars can be resolved out of the five that exist in the system. Future studies with advanced telescopes and imaging techniques will help scientists confirm the number of stars in this system. Studying the Acrux system is challenging because it contains multiple closely orbiting stars. While new technology is improving our understanding, fully unraveling its complexity remains difficult.

STAR CHARACTERISTICS

Star Name	Acrux
Rank	13th brightest star in the night sky
Pronunciation	A-krucks
Star Name Meaning	Combination of "A" for Alpha (brightest) and "Crux" the constellation
Alternative Names	Alpha Crucis
Constellation	Crux the Southern Cross
Location (Right Ascension, Declination)	$12^h\ 26^m\ 35^s$ $-63°\ 05'\ 56''$

Distance	321 light-years	
Magnitude	Apparent: +0.76	Absolute: -3.77
System	Multi-star system: Acrux Aa—main sequence blue giant star Acrux Ab—main sequence blue giant star Acrux B—possible main sequence star Acrux Ca—main sequence blue giant star Acrux Cb—companion star	
Temperature*	28,000 Kelvin	
Spectral Class*	B-type star	
Visible From	Northern Hemisphere: 0° to +27°N	Southern Hemisphere: 0° to -90°S
Compared to Our Sun*	Mass: 17 Solar Masses	Luminosity: 31,000 times brighter than the sun
	Radius: 7 Solar Radii	
Exoplanets	None detected	
Eventual Fate	Acrux Aa and Acrux B are both large enough to explode into supernovae, then evolve into either a black hole or neutron star.	

Interesting Facts	+ This star is classified as a spectroscopic binary, meaning the two stars cannot be resolved in a telescope, but are detected by analyzing the spectra of a star and looking for tiny shifts in its brightness.
+ Acrux is the brightest star and southernmost star in the constellation of Crux, the Southern Cross.
+ Acrux is only visible in the northern hemisphere from latitudes 27 degrees to the equator.
+ Acrux is seen on multiple flags, including Australia, New Zealand, Samoa, and Papua New Guinea. It also appears on the flag of Brazil, and it represents the state of São Paulo.
+ Acrux is the southernmost first-magnitude star.
+ At latitude -27° south, Acrux would be a circumpolar star, which means it can be seen every night of the year.
+ The Cassini-Huygens spacecraft that orbited Saturn took an image of the Acrux system in 2008. It was able to resolve three of the five stars in the system.

*Primary star Acrux Aa

Aldebaran (Taurus)—The Follower

Aldebaran is an orange giant star and is the brightest star in the constellation Taurus the Bull. It is approximately 65 light-years away from Earth. Its luminosity, or brightness, is over 500 times that of our sun and is classified as a K-type giant star. It is a key reference point for observers to locate other celestial objects. Aldebaran has been referenced in various cultures and has had many different names throughout history. Its name, "Aldebaran," is derived from the Arabic phrase "al-dabarān," which means "the follower." This name reflects its position as it appears to follow the Pleiades Star cluster (Seven Sisters) across the night sky.

Aldebaran is classified as a K-type giant star, which means it is in the later stages of its stellar evolution. This star has been heavily studied because it is an important object for examining the late stages of stellar evolution in giant-type stars. Some studies have suggested this may be a binary star system with a faint companion red dwarf star; however, this has been difficult to confirm since Aldebaran is so bright.

In order to understand why Aldebaran and our sun are so different, we have to look at the life cycle of a star. Each of these stars are at a different stage in their lives. Our sun is still in the main sequence phase, where the forces of gravity and fusion are balanced. It's a steady star. Aldebaran, on the other hand, has evolved off the main sequence band. This indicates that a star can no longer fuse hydrogen at its core. Instead, hydrogen fusion occurs in the outer layers, causing the star to expand. Helium fusion now occurs in the core, and the surface temperature decreases. But even though the star's surface temperature decreases, it glows more because of its enormous size.

This orange-hued star is easily visible to the naked eye and has been observed and referenced by many ancient cultures. This star looks like it could be a part of the Hyades Star cluster, however it is located 65 light-years away, while the Hyades Star Cluster is estimated to be 150 light-years away. Aldebaran holds cultural significance in different mythologies and has been a subject of interest for astronomers and sky gazers throughout history.

STAR CHARACTERISTICS

Star Name	Aldebaran	
Rank	14th brightest star in the night sky	
Pronunciation	al-DEB-ah-ran	
Star Name Meaning	Arabic for "the follower"	
Alternative Names	Alpha Tauri	
Constellation	Taurus the Bull	
Location (Right Ascension, Declination)	$4^h\ 35^m\ 55^s$ +16° 30′ 33″	
Distance	Approx. 65 light-years	
Magnitude	Apparent: +0.86	Absolute: -0.64
System	Single star system: Aldebaran—orange giant star	
Temperature	3,900 Kelvin	
Spectral Class	K-type star	
Visible From	Northern Hemisphere: 0° to +90°N	Southern Hemisphere: 0° to -65°S
Compared to Our Sun	Mass: 1.16 Solar Masses	Luminosity: Approx. 430 times brighter than the sun
	Radius: 45 Solar Radii	
Exoplanets	Aldebaran b is a possible exoplanet candidate (awaiting confirmation)	
Eventual Fate	Planetary nebula, then white dwarf	

Interesting Facts	✦ Aldebaran is a variable star, which means its brightness changes over time. ✦ Aldebaran has exhausted the hydrogen in its core. Scientists have observed an abundance of carbon, oxygen, and nitrogen in its photosphere. ✦ This star looks like it could be a part of the Hyades Star cluster, however it is located 65 light-years away, while the Hyades Star Cluster is estimated to be 150 light-years away. ✦ Pioneer 10, a NASA space probe launched in 1972, will come in contact with the Aldebaran system in two million years, as it is headed in that direction of the sky. Pioneer 10 was the first artificial satellite to achieve the escape velocity to leave the solar system. ✦ Aldebaran may be a possible binary system; however, the faint red dwarf star is difficult to detect since the main star is so bright.

Antares (Scorpius)—Heart of the Scorpion

Antares, the heart of the scorpion, is a bright red star whose color rivals that of the planet Mars. In fact, that is where its name comes from. Its name is derived from the Greek prefix "Anti-" meaning "against" and "Ares" meaning "Mars." When combined it means "rival of Mars" because its deep red color is very similar to that of the red planet Mars. Its distance is 600 light-years from Earth.

The Antares system is a binary system: Antares A and Antares B. Antares A, the largest of the pair, is classified as a red supergiant, so it is a star nearing the end of its life. In the supergiant phase, heavier elements such as carbon, oxygen, nitrogen and neon are being fused. Eventually iron will begin to form inside the star, and in time will cause the collapse of the star into a supernova. Antares B is a bluish-white main sequence star and is the much smaller of the two stars in this system. This star has fascinated astronomers because it appears to have a greenish tint to its color, which is most likely a contrast effect caused by its companion supergiant star. There are no green stars in existence (see FAQ section for more details). At a mass of 4 solar masses, Antares B does not meet the threshold needed to become a supernova. Instead, it is expected to evolve into a red giant, then a planetary nebula, and eventually a white dwarf. These two stars are estimated to be 529 astronomical units apart.

The radius of Antares A is approximately 833 times that of our sun. Let's put it this way: if Antares A were to be placed in our solar system, the edge of the star would lie between the orbits of Mars and Jupiter. It is speculated that Antares A is only 12 million years old, which is fairly young compared to other stars in our local system. This star does have the potential to explode into a supernova in the next few hundred thousand years. If this were to happen, it could be as bright as the moon and be visible during the daytime.

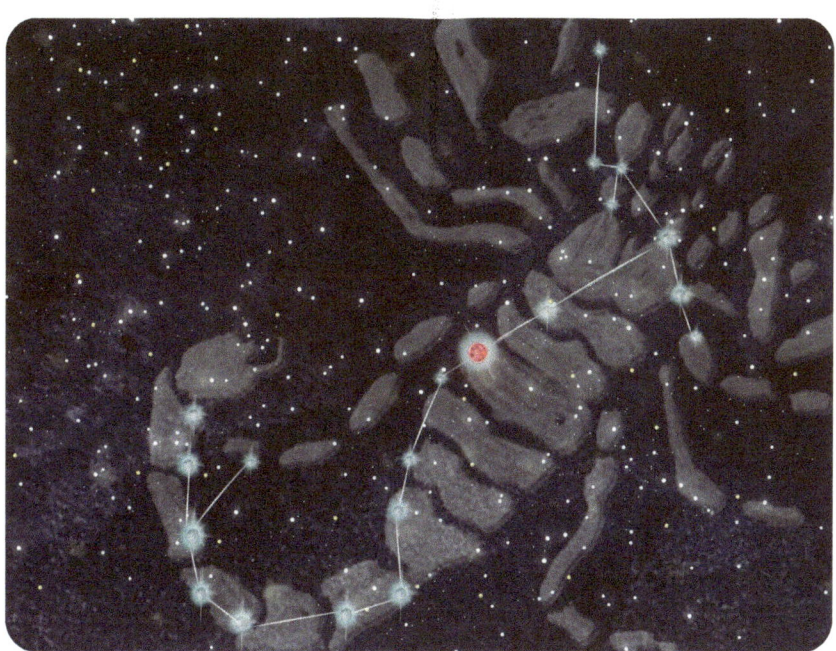

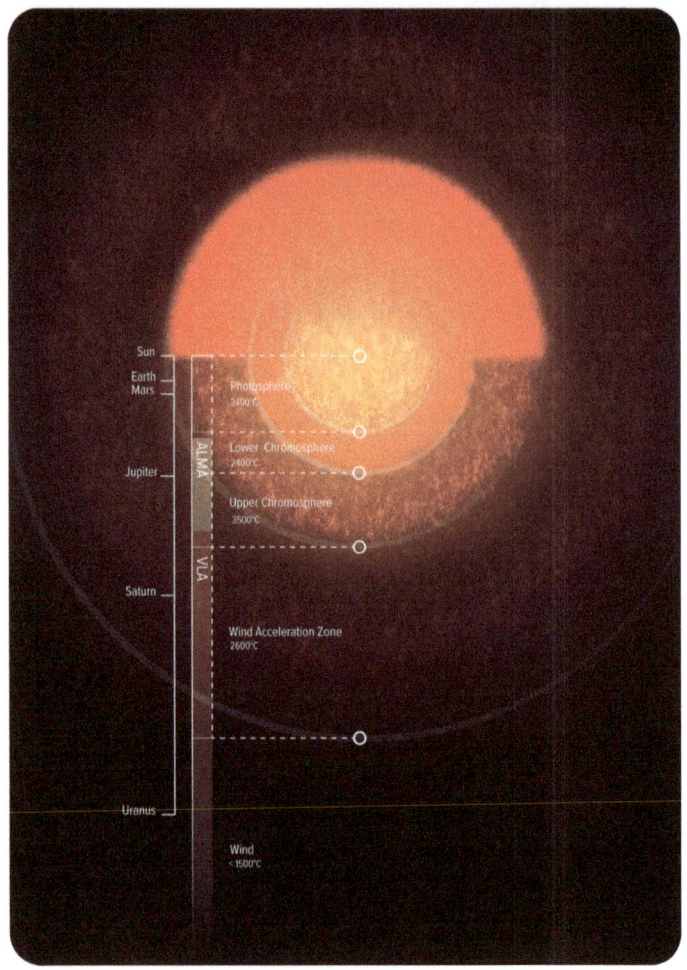

The atmosphere of Antares compared to our own solar system.

STAR CHARACTERISTICS

Star Name	Antares
Rank	15th brightest star in the night sky
Pronunciation	an-TAIR-ease
Star Name Meaning	Greek prefix "anti-" meaning "against" and "Ares" meaning "Mars"
Alternative Names	Alpha Scorpii, Heart of the Scorpion
Constellation	Scorpius the Scorpion

Location (Right Ascension, Declination)	16ʰ 29ᵐ 24ˢ -26° 25' 55"
Distance	600 light-years
Magnitude	Apparent: +1.06	Absolute: -5.28
System	Binary star system: Antares A—red supergiant star Antares B—main sequence blue giant star
Temperature	Antares A—3,600 Kelvin Antares B—18,500 Kelvin
Spectral Class*	M-type star
Visible From	Northern Hemisphere: 0° to +40°N	Southern Hemisphere: 0° to -90°S
Compared to Our Sun*	Mass: 12 Solar Masses	Luminosity: 75,000 times brighter than the sun
	Radius: 680–700 Solar Radii	
Exoplanets	None detected
Eventual Fate*	Supernova, then black hole or neutron star
Interesting Facts	✦ Antares has been referenced in various cultures and has had different names throughout history. Its name, "Antares," is derived from the Ancient Greek word "Anti-Ares," which means "rival of Mars." The star's reddish appearance is similar to the planet Mars, and it was named for its rivalry in brightness and color. ✦ Antares A is a variable star, so its brightness fluctuates over time. ✦ The name of this star in Mesopotamian astronomy is "heart of Scorpion" and it was associated with the goddess Lisin. ✦ Antares is one of the few stars that has been directly imaged, allowing scientists to get a better understanding of the surface of Antares. It was first imaged in 2017 by the Very Large Telescope, a ground-based facility located in Chile and operated by the European Space Agency (ESO).

*Primary star Antares A

Spica (Virgo)—Ear of Wheat Star

Spica is located in Virgo, one of the zodiac constellations. Its name, "Spica," comes from the Latin word for "ear of grain" or "spike," alluding to its link with the constellation Virgo, which is sometimes portrayed as a woman carrying a sheaf of wheat. Spica's brightness and distinctive color make it a popular star to observe and identify in the night sky. It's a notable member of a zodiac constellation and is often used as a reference point for stargazers and astronomers. Spica is often used for celestial navigation and orientation in the night sky. It is part of the asterism known as the "Spring Triangle," along with Arcturus and Regulus, which helps observers locate various celestial objects.

Spica is a binary star system made up of two massive, blue main sequence stars that orbit a shared center of mass. The Spica star system is around 250 light-years from Earth, making it a reasonably close neighbor. These two stars have a four-day orbital period. Both stars' shapes are influenced by their rapid orbits. Due to their tremendous gravitational attraction and quick orbit, both stars are most likely egg-shaped. These two stars are estimated to be 0.1 astronomical units away from each other. For reference, the sun and the planet Mercury are 0.3 astronomical units away. So Spica A and Spica B are closer than Mercury and the sun!

STAR CHARACTERISTICS

Star Name	Spica
Rank	16th brightest star in the night sky
Pronunciation	SPY-ka or SPEE-ka
Star Name Meaning	Latin word for "ear of grain" or "spike"
Alternative Names	Alpha Virginis
Constellation	Virgo the Virgin
Location (Right Ascension, Declination)	$13^h\ 25^m\ 12^s$ -11° 09' 41"

Distance	250 light-years	
Magnitude	Apparent: +0.98	Absolute: -3.55
System	Binary star system: Spica A—main sequence blue giant star Spica B—main sequence blue giant star	
Temperature*	25,000 Kelvin	
Spectral Class*	B-type star	
Visible From	Northern Hemisphere: 0° to +80°N	Southern Hemisphere: 0° to -80°S
Compared to Our Sun*	Mass: 10 Solar Masses	Luminosity: Approx. 2,250 times brighter than the sun
	Radius: 7 Solar Radii	
Exoplanets	None detected	
Eventual Fate*	Supergiant, supernova, then black hole or neutron star	
Interesting Facts	✦ Spica is a variable star, which means its brightest changes over time, and this is because it is an eclipsing binary system. An eclipsing binary when the orbital plane of two stars is oriented in such a way that, as seen from Earth, one star periodically passes in front of the other, causing eclipses. This causes a change in brightness. ✦ The two stars in this system have an egg-like shape due to their close orbit, completing a revolution around each other every four days. This near proximity causes their shapes to be distorted by gravitational forces. ✦ Spica A is estimated to be 12 solar masses, while Spica B is estimated to be 7 solar masses. They are both the same spectral class. ✦ Since Spica is only two degrees away from the ecliptic, planets can sometimes pass in front of the star, and this is called *occultation*.	

*Primary star Spica A

Pollux (Gemini)—Red Giant of Gemini

Pollux, also called Beta Geminorum, is the brightest star in the constellation Gemini. Castor, the star next to Pollux in the sky, forms a dazzling duo that is plainly visible in the night sky. Pollux is around 34 light-years away from Earth, making it one of our closer companions in the local stellar neighborhood. It is approximately twice the mass of the sun and nine times its radius. Pollux has depleted the hydrogen in its core and developed into a massive star. It is currently fusing helium and other heavier elements in its core.

An exoplanet called Pollux b (also nicknamed Thestias) has been detected around this star. Pollux b was identified in 2006 using the radial velocity method. The radial velocity method involves observing slight changes in the motion of a star caused by the gravitational influence of an orbiting planet. Pollux b is a gas giant planet with a mass of 2.3 times that of Jupiter and orbits Pollux at a distance of about 1.64 astronomical units with an orbital period of approximately 590 days.

Pollux is one of the twins symbolized by the Gemini constellation. Pollux and Castor are renowned in Greek mythology as the Dioscuri, the twin sons of Zeus (or Jupiter in Roman mythology) and Spartan queen Leda. In some versions of the narrative, Pollux is Zeus' son, making him immortal, whereas Castor is the son of Tyndareus, Sparta's mortal king. The twins were inseparable and experienced numerous adventures, including accompanying the Argonauts on their hunt for the Golden Fleece. The twins were also thought to protect sailors and were frequently invoked for safety on maritime voyages. When Castor was killed, Pollux, who is immortal, pleaded with Zeus to let him share his immortality with his brother. Zeus placed them both in the constellation Gemini, allowing them to stay together forever.

Pollux is easily visible to the naked eye due to its brightness. It is best observed during the winter months in the northern hemisphere or the summer months in the southern hemisphere. Pollux and Castor are right next to each other in the Gemini, forming a noticeable pair of bright stars. Pollux is the brighter of the two and is orange-hued, while Castor is slightly dimmer and bluish white in color.

STAR CHARACTERISTICS

Star Name	Pollux	
Rank	17th brightest star in the night sky	
Pronunciation	POL-lucks	
Star Name Meaning	Pollux, twin brother of Castor, figures in Greek mythology	
Alternative Names	Beta Geminorum	
Constellation	Gemini the Twins	
Location (Right Ascension, Declination)	$07^h\,45^m\,19^s$ $+28°\,01'\,34''$	
Distance	33 light-years	
Magnitude	Apparent: +1.14	Absolute: +1.08
System	Single star system: Pollux—orange giant star	
Temperature	4,500 Kelvin	
Spectral Class	K-type star	
Visible From	Northern Hemisphere: 0° to +90°N	Southern Hemisphere: 0° to -60°S
Compared to Our Sun	Mass: 2 Solar Masses	Luminosity: 32 times brighter than the sun
	Radius: 9 Solar Radii	
Exoplanets	Pollux b (gas giant)	
Eventual Fate	Planetary nebula, then white dwarf	

Interesting Facts	✦ Despite its beta designation, Pollux is the brighter star of the twins. The beta prefix is used for the second-brightest star in a constellation.
	✦ Pollux is cooler than the sun, but brighter. This is because Pollux is no longer a main sequence star. It has evolved into a red giant and has grown in size from its original size. Pollux is nine times wider than our sun.
	✦ Pollux is an example of what our sun will eventually become in its later years—a giant star that grows in size as it runs out of hydrogen in the core.
	✦ Pollux and Castor appear to have different colors when viewed at night. Pollux is orange-hued, while Castor is bluish in appearance.

Fomalhaut (Piscis Austrinus)— The Loneliest Star

Fomalhaut, also known as Alpha Piscis Austrini, is one of the brightest stars in the night sky and is the most prominent star in the constellation Piscis Austrinus (the Southern Fish). Sometimes dubbed the Loneliest Star, it is the only first-magnitude star in this region of the sky. To the naked eye it appears as one star, but upon closer inspection, it forms a triple star system. Fomalhaut A is the primary star, while the stars TW Piscis Austrini (Fomalhaut B) and LP 876-10 (Fomalhaut C), are both much fainter and located at greater distances from the primary star. This star system is located 25 light-years away, making it a close stellar neighbor.

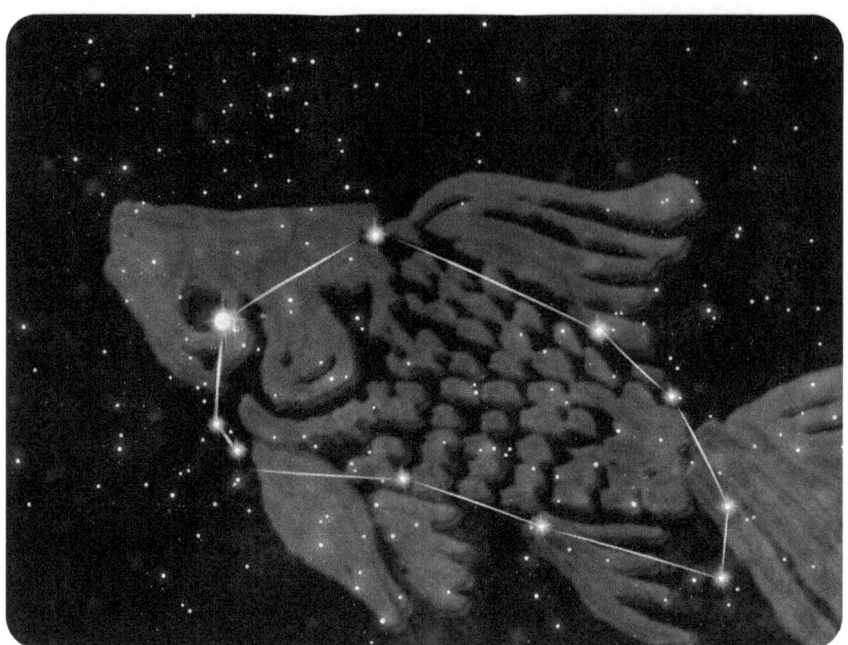

Fomalhaut can be found using the Great Square of Pegasus. Find the two stars on the western side of the Square and draw a line through them and go south through Aquarius about 45 degrees. There are no bright stars along the way to guide toward Fomalhaut. It is the only bright star in a group of dim ones, which is how it earned its name: the "Loneliest Star."

Fomalhaut is best viewed from the southern hemisphere where it is visible from August until November. It can be viewed in the lower latitudes in the northern hemisphere, but it does not rise high in the sky and is only visible for a maximum of eight hours. In the northern hemisphere, it is most prominent in the evening sky during autumn months, hence the additional nickname called the "Autumn Star."

Fomalhaut was the first star system to have an extrasolar planet candidate imaged at visible wavelengths. The Hubble Space Telescope was the first to image this exoplanet and was named Fomalhaut b. However, more recent research and new observations of this star show that Fomalhaut b is not a planet, but rather a growing patch of debris from a catastrophic planetesimal collision. In 2020, the

NASA Exoplanet Archive formally removed Fomalhaut b from its list of exoplanet candidates.

Fomalhaut A star is very similar in structure to our own solar system and the star Vega. All of these systems have debris disks surrounding them. Fomalhaut A has been confirmed to have multiple debris disks. There is an inner, middle, and outer debris ring that surrounds the main star. These debris disks are similar to our own asteroid belt and Kuiper Belt.

The key to finding Fomalhaut is to use the Great Square of Pegasus asterism. In this photo, look for the giant square shape. Use the two stars on the right side of the square to point down toward Fomalhaut (not pictured). Remember, there are no other bright stars near Fomalhaut.

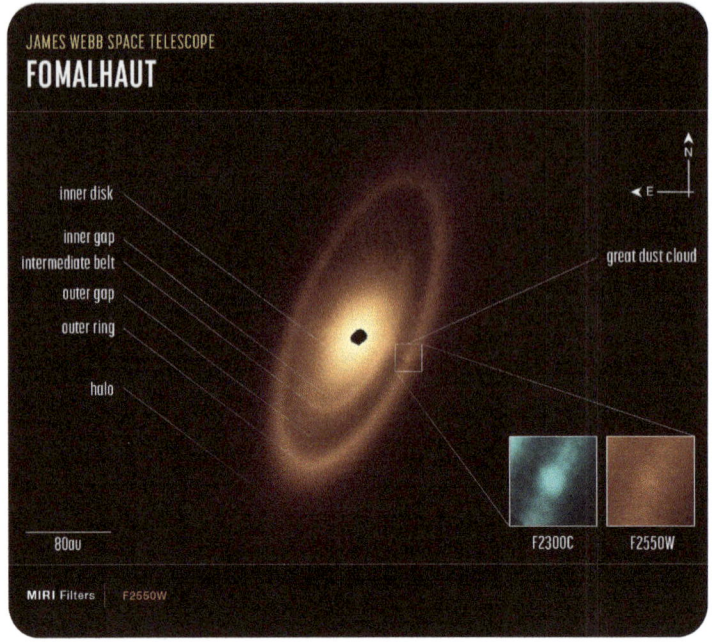

Data reveals that Fomalhaut has three dusty debris rings, similar to our own asteroid belt and Kuiper Belt. There are three rings: an inner, middle, and outer ring. These debris rings, known as circumstellar debris disks, produce excess infrared radiation and are made up of gas, dust, planetesimals, and asteroids.

STAR CHARACTERISTICS

Star Name	Fomalhaut
Rank	18th brightest star in the night sky
Pronunciation	FO-mal-oh or FOH-mal-owt
Star Name Meaning	Arabic name meaning "mouth of the whale"
Alternative Names	Alpha Piscis Austrini
Constellation	Piscis Austrinus the Southern Fish
Location (Right Ascension, Declination)	$22^h\ 57^m\ 39^s$ $-29°\ 37'\ 20''$
Distance	25 light-years

Magnitude	Apparent: +1.16	Absolute: +1.72
System	Triple star system: Fomalhaut A—white main sequence star Fomalhaut B—orange main sequence star Fomalhaut C—main sequence red dwarf star	
Temperature✱	8,500 Kelvin	
Spectral Class✱	A-type star	
Visible From	Northern Hemisphere: 0° to +55°N	Southern Hemisphere: 0° to -90°S
Compared to Our Sun✱	Mass: 2 Solar Masses	Luminosity: 16 times brighter than the sun
	Radius: 2 Solar Radii	
Exoplanets	None detected	
Eventual Fate✱	Giant star, planetary nebula, then white dwarf	
Interesting Facts	✦ Fomalhaut b was once believed to have an exoplanet, but was ultimately identified as a series of debris disks surrounding the star. These circumstellar disks contain dust and gas much like our own solar system. ✦ Fomalhaut is the only bright star in its region of the sky, giving it the nickname the "Lonely Star." ✦ Like many stars, Fomalhaut has many nicknames. In the northern hemisphere it has been dubbed the "Autumn Star" because it is one of the only bright stars in the night sky during this time of year. ✦ The main star in this system, Fomalhaut A, is brighter, hotter, and larger than the sun. It is about double the mass and size of the sun.	

✱Primary star Fomalhaut A

Deneb (Cygnus)—Tail of the Swan

Deneb is the brightest star in the Cygnus constellation and the 19th brightest star at night. It is a bluish white in color and classified as a first-magnitude supergiant star. Deneb's mass is estimated to be 19 times that of the sun. This star is actively expanding as hydrogen is fused in its outer layers. This star is substantially larger than the sun and is estimated to be 200 times the size of the sun and 200,000 times brighter. Its temperature range is significantly hotter than the sun, which contributes to its overall brightness, along with its large size. Deneb is also categorized as a variable star, so its brightness changes over time.

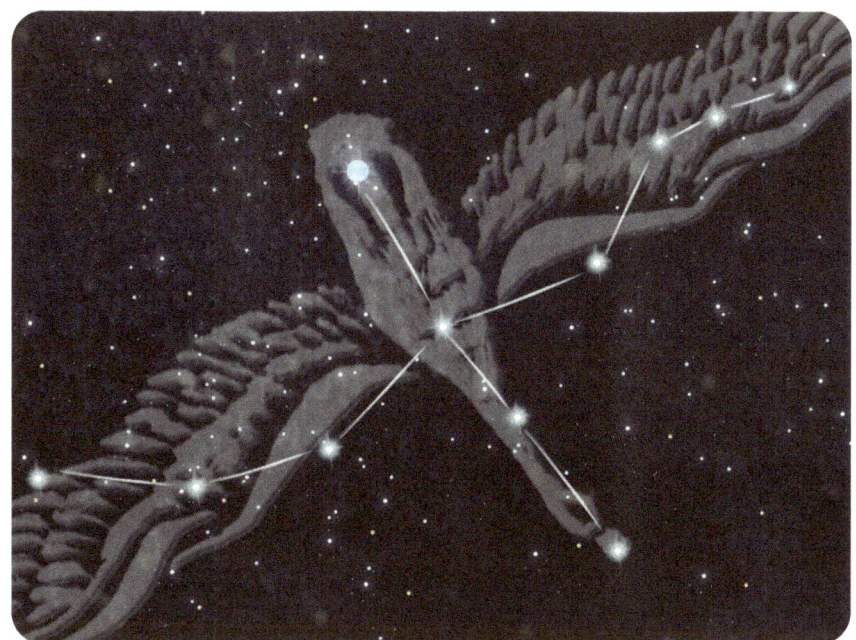

Deneb is visible in both the northern and southern hemispheres. In the northern hemisphere, it is visible at all latitudes in the summer months. In the southern hemisphere, Deneb can be seen from the equator down 40-45 degrees latitude during the winter season. The first step in identifying Deneb in the night sky is to find the Summer Triangle asterism. The dimmest of the three stars in the Summer Triangle is Deneb.

Deneb is an important star to recognize and is useful when trying to find other asterisms and constellations. Deneb is part of two distinct asterisms. One is the Northern Cross, a pattern within the constellation of Cygnus. The other asterism is the Summer Triangle, and it contains three stars: Vega, Altair, and Deneb.

The three brightest stars in this photo connect to form the Summer Triangle. The star furthest to the left is Deneb. Deneb also makes up the Northern Cross asterism. Can you find the cross pattern using Deneb?

Astronomers have struggled to determine the distance to Deneb. It is predicted to be between 1,500 and 2,600 light-years away, which is a quite broad range in terms of distance. When the European Space Agency's Hipparcos mission measured the distance to Deneb, there was a large margin of error. The Hipparcos mission aimed to determine the velocity and distance between stars in the sky. This information would later be used to chart the stars in the night sky. The GAIA mission (short for Global Astrometric Interferometer for Astrophysics) succeeded Hipparcos and continued to perform the same mission but with greater precision. Nevertheless, Deneb is simply too bright for GAIA to measure. This bright star saturates the telescope's sensor, preventing it from accurately detecting distance. The brightest magnitude that GAIA can measure is 1.71, and Deneb is brighter at 1.25 magnitude.

	Hipparcos Stats	GAIA Stats
Commissioned By	European Space Agency	
In Service	1989 to 1993	2013 to 2025
Insignia		

Celestial Objects

The North American Nebula is a celestial object that lies close to Deneb in the night sky. A nebula is a cloud of gas and dust and is often named after its shape and this one resembles the continent of North America.

Mythology

The fable of Zhinü and Niulang is a well-known and popular story in Chinese mythology, frequently connected with the Qixi Festival, also known as the Double Seventh Festival. This romance story has been passed down over the centuries in many Asian cultures. In this love story, the Weaver Girl, Zhinü, is a celestial being and the seventh daughter of the Jade Emperor, who is the ruler of Heaven. She is skilled at weaving beautiful, intricate fabrics. Niulang, the cowherd, is a kind and hardworking mortal man. The lovers are separated after their love becomes all consuming, but are allowed to reunite once a year. The star Vega represents Zhinü, and the star Altair represents Niulang. Deneb is the star that helps guide the lovers to reunite. The beauty of this cherished love story is reflected in the night sky year after year (see also Vega and Altair sections).

STAR CHARACTERISTICS

Star Name	Deneb	
Rank	19th brightest star in the night sky	
Pronunciation	DEN-ebb	
Star Name Meaning	Arabic term meaning "tail"	
Alternative Names	Alpha Cygni	
Constellation	Cygnus the Swan	
Location (Right Ascension, Declination)	$20^h\ 41^m\ 25^s$ +45° 16' 49"	
Distance	1,500–2,600 light-years (Exact distance unclear with current measurements)	
Magnitude	Apparent: +1.25	Absolute: -8.38
System	Single star system: Deneb—white supergiant star	

Temperature	8,500 Kelvin	
Spectral Class	A-type star	
Visible From	Northern Hemisphere: 0° to +90°N	Southern Hemisphere: 0° to -40°S
Compared to Our Sun	Mass: 19 Solar Masses	Luminosity: 200,000 times brighter than the sun
	Radius: 200 Solar Radii	
Exoplanets	None detected	
Eventual Fate	Supernova, then neutron star or black hole	
Interesting Facts	<ul><li>Deneb will be the pole star in 9800 AD. It will be 7 degrees off the North Celestial Pole.</li><li>If Deneb were to be placed inside our own solar system, its boundaries would extend to the orbit of Earth.</li><li>Deneb is the dimmest star of the Summer Triangle asterism. The other two stars, Vega and Altair, are slightly brighter than Deneb. However, all three are easy to point out in the sky due to their brightness and pattern.</li></ul>	

Mimosa (Crux)—Binary Star System

Mimosa, also known as Beta Crucis, is the second-brightest star in the constellation Crux the Southern Cross and is found in the southern hemisphere. It is a binary star system that is roughly 280 light-years from Earth. The two stars, Beta Crucis A and Beta Crucis B are both blue main sequence stars. This system is a spectroscopic binary, which means the two stars cannot be resolved in a telescope. The companion star that cannot be seen directly and only through detailed analysis of this system's light. Beta Crucis A is the largest of the two stars at 16 solar masses, while Beta Crucis B is smaller at 10 solar masses. It takes five years for these two stars to orbit each other.

STAR CHARACTERISTICS

Star Name	Mimosa
Rank	20th brightest star in the night sky
Pronunciation	mim-OH-sah
Star Name Meaning	Latin meaning "actor"; mimosa also represents a flower. Origin of this star name is unclear
Alternative Names	Beta Crucis
Constellation	Crux the Southern Cross
Location (Right Ascension, Declination)	$12^h\ 47^m\ 43^s$ -59° 41' 19"
Distance	280 light-years
Magnitude	Apparent: +1.25	Absolute: -3.92

System	Binary star system: Beta Crucis A—main sequence blue giant star Beta Crucis B—main sequence blue giant star	
Temperature*	27,000 Kelvin	
Spectral Class*	B-type star	
Visible From	Northern Hemisphere: 0° to +20°N	Southern Hemisphere: 0° to -90°S
Compared to Our Sun*	Mass: 16 Solar Masses	Luminosity: 34,000 times brighter than the sun
	Radius: 8 Solar Radii	
Exoplanets	None detected	
Eventual Fate*	Supergiant, supernova, then black hole or neutron star	
Interesting Facts	✦ These stars are classified as a spectroscopic binary. ✦ The two stars in this system orbit each other every five years. ✦ The Jewel Box cluster lies to the east of Mimosa. It is a favorite object of many astrophotographers to capture. ✦ The Coalsack Nebula, a darker region of the sky in which light does not pass through, is located directly south of Mimosa. ✦ Beta Crucis A is a beta cepheid variable star, which means it has regular and predictable pulsations and its brightness varies over time.	

*Primary star Beta Crucis A

Regulus (Leo)—Heart of the Lion

Regulus, also known as Alpha Leonis, is the brightest star in the constellation Leo. It marks the heart of the lion in the constellation's traditional depiction. Regulus is a multiple star system, consisting of at least four stars. The primary component, Regulus A, is a blue-white main sequence star and is the largest star of the system. Regulus A has a close white dwarf binary companion. Regulus B and C form a wider pair orbiting the primary pair at a greater distance. The entire star system is located approximately 79 light-years away from Earth.

The main star, Regulus A, has a surface temperature of around 12,000 K, making it much hotter than the sun. It is about 3.5 times more massive than the sun and 300 times more luminous. Regulus A has an exceptionally high rotational velocity, making one rotation approximately every sixteen hours. This rapid rotation causes it to be significantly flattened at the poles. This rapid rotation also causes the star to be hotter at the poles than at the equator, an effect known as *gravity darkening* and *gravity lightening*. This star's poles are brighter than the equator due to its rapid spin. Thus, the poles are "gravity brightened" and the equator is "gravity darkened."

STAR CHARACTERISTICS

Star Name	Regulus
Rank	21st brightest star in the night sky
Pronunciation	REG-you-luss
Star Name Meaning	Latin for "little king"
Alternative Names	Alpha Leonis
Constellation	Leo the Lion
Location (Right Ascension, Declination)	$10^h\ 08^m\ 22^s$ $+11°\ 58'\ 02''$
Distance	79 light-years

Magnitude	Apparent: +1.4	Absolute: -0.57
System	Quadruple star system: Regulus Aa—main sequence blue star Regulus Ab—white dwarf star Regulus B—main sequence orange dwarf star Regulus C—main sequence red dwarf star	
Temperature*	12,000 Kelvin	
Spectral Class*	B-type star	
Visible From	Northern Hemisphere: 0° to +90°N	Southern Hemisphere: 0° to -65°S
Compared to Our Sun*	Mass: 3.8 Solar Masses	Luminosity: 300 times brighter than the sun
	Radius: 4 Solar Radii	
Exoplanets	None detected	
Eventual Fate*	Giant star, planetary nebula, then white dwarf	

Interesting Facts	✦ Regulus is an egg-shaped star because of its rapid rotational speed in comparison to other stars.
	✦ The main star, Regulus A, is a spectroscopic binary star. This means that the second star can only be detected by analyzing the light coming from the star. Scientists study spectroscopic binaries to learn about its properties since it is invisible to the unaided eye.
	✦ The companion stars of Regulus A are too dim to see with the unaided eye. A powerful telescope is needed to resolve the two stars orbiting Regulus A.
	✦ Regulus lies very close to the ecliptic, which is the path of the sun. Sometimes it can be blocked by the moon or planets. This type of event is called an occultation.
	✦ Regulus is the brightest star in the Leo constellation. While the namesake of this constellation is attributed to the Greeks, many cultures before the Greeks recognized this star pattern to be a lion.

*Primary star Regulus Aa

Adhara (Canis Major)— Binary Star System

Adhara, also known as Epsilon Canis Majoris, is the night sky's 22nd brightest star and the second brightest in the Canis Major constellation. It is a binary star system that is sometimes overlooked as a bright star because Sirius outshines the majority of stars in this constellation. The name "Adhara" is derived from the Arabic "Al 'Adhārā," which means "the virgins."

The main star of this system, Adhara A, is a blue, main sequence star that is estimated to be 430 light-years away. This star is unique because it is one of the brightest sources of ultraviolet radiation in the night sky. If human eyes were able to detect ultraviolet radiation instead of visible light, Adhara would be the brightest star in the night sky. Adhara B is a small star located 900 astronomical units from Adhara A. These two stars complete an orbit around each other approximately every 7,500 years. A powerful telescope is required to distinguish them as separate points of light.

STAR CHARACTERISTICS

Star Name	Adhara
Rank	22nd brightest star in the night sky
Pronunciation	ad-HAR-a
Star Name Meaning	Arabic word meaning "maidens" or "virgins"
Alternative Names	Epsilon Canis Majoris
Constellation	Canis Major the Great Dog
Location (Right Ascension, Declination)	$6^h\ 58^m\ 38^s$ $-28°\ 58'\ 19''$
Distance	430 light-years
Magnitude	Apparent: +1.5 — Absolute: -4.8

System	Binary star system: Adhara A—blue main sequence star Adhara B—companion star	
Temperature*	22,900 Kelvin	
Spectral Class*	B-type star	
Visible From	Northern Hemisphere: 0° to +60°N	Southern Hemisphere: 0° to -90°S
Compared to Our Sun*	Mass: 12 Solar Masses	Luminosity: 38,700 times brighter than the sun
	Radius: Approx. 14 Solar Radii	
Exoplanets	None detected	
Eventual Fate*	Supergiant, supernova, then black hole or neutron star	
Interesting Facts	✦ Adhara A outshines its companion, Adhara B, by 250 times. ✦ Adhara is the brightest extreme ultraviolet source in the night sky. ✦ The origin of this star name comes from the Arabic term meaning "the maidens" or "the virgins," but it is unclear as to why it was named after this term. The answer has been lost to history. ✦ If Adhara were as close to Earth as Sirius is (8 light-years away), it would outshine Venus in the night sky!	

*Primary star Adhara A

Shaula (Scorpius)— The Scorpion's Stinger

Shaula, also known as Lambda Scorpii, is a prominent star in the constellation Scorpius. It is known for its brightness and its position as part of the "stinger" of the Scorpion. It is a triple star system located about 570 light-years away from Earth.

The primary star in this system is Shaula Aa, and secondary star Shaula B are both blue, main sequence stars. The smallest star of this system, Shaula Ab, is smaller in size and orbits the main star Shaula Aa. It is unclear as to what type of star it is. Scientists have speculated that this star could be a protostar, a baby star not yet undergoing fusion.

Lesath is the other star that represents the "stinger" of the Scorpion. These two stars are separated by a distance of 5 light-years and together are nicknamed "Cat Eyes." Messier 6 (Butterfly Cluster) and Messier 7 (Ptolemy Cluster) are star clusters located near Shaula and Lesath. Both are visible with the unaided eye.

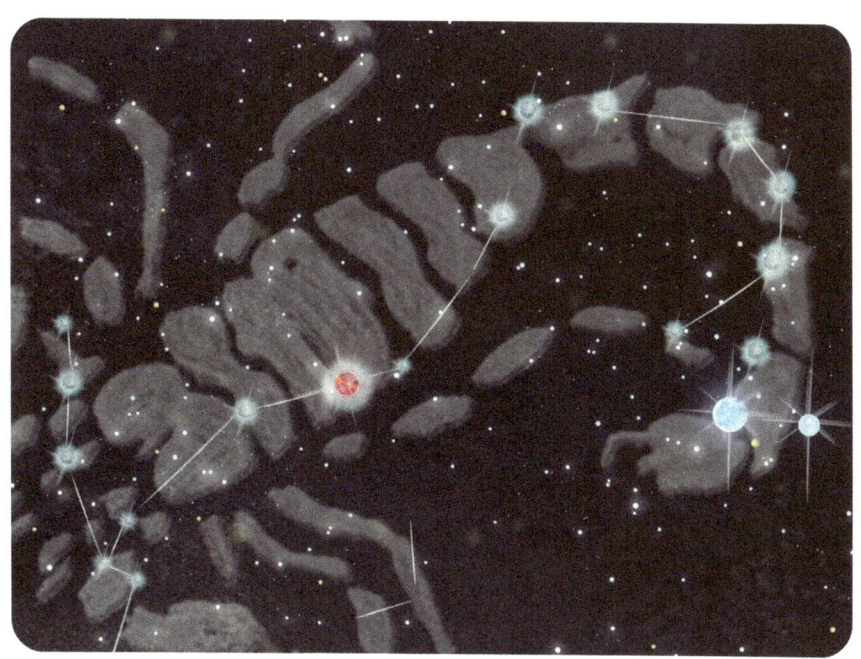

STAR CHARACTERISTICS

Star Name	Shaula	
Rank	23rd brightest star in the night sky	
Pronunciation	SHOWL-a	
Star Name Meaning	Arabic meaning "raised tail"	
Alternative Names	Lambda Scorpii	
Constellation	Scorpius	
Location (Right Ascension, Declination)	$17^h 33^m 36^s$ $-37°\,06'\,13''$	
Distance	570 light-years	
Magnitude	Apparent: +1.62	Absolute: -3.7

System	Triple star system: Shaula Aa—blue main sequence star Shaula Ab—possible pre-main sequence star Shaula B—blue main sequence star
Temperature*	25,000 Kelvin
Spectral Class*	B-type star
Visible From	Northern Hemisphere: 0° to +40°N / Southern Hemisphere: 0° to -90°S
Compared to Our Sun*	Mass: 10 Solar Masses / Luminosity: 35,000 times brighter than the sun Radius: 8 Solar Radii
Exoplanets	None detected
Eventual Fate*	Supergiant, supernova, then black hole or neutron star
Interesting Facts	✦ This star is in the tail of the Scorpius, and it forms part of the star pattern that represents the "stinger" of the scorpion. ✦ This star was first recorded in the Babylonian Star Catalogs. ✦ The smallest star in this system, Star Ab, is not a true star yet. It has enough mass in the surrounding gas and dust, but the process of fusion has not yet ignited in the core. ✦ Shaula is a beta cepheid star, a type of variable star. This means the surface of the star pulsates, which can make the brightest of the star fluctuate. ✦ Shaula is the second-brightest star in Scorpius. Antares is the brightest.

*Primary star Shaula Aa

Castor (Gemini)—Six Star System

Castor, or Alpha Geminorum, is the second-brightest star in the Gemini constellation (Pollux is the brightest). It is a complex star system, notable for its six stars organized into three binary pairs. In Greek mythology, Castor and Pollux (the Gemini twin stars) are associated with the legendary twins Castor and Pollux, sons of Leda and Argonauts.

Both Castor A and Castor B are both spectroscopic binary stars that consist of an A-type, white main sequence star (Aa, Ba) and a red dwarf (Ab, Bb). The Castor C binary pair consists of two red dwarf stars (Ca, Cb). To the Earth-bound observer, Castor appears to be a single star to the naked eye, but when viewed through a small telescope, it is possible to reveal two of the six stars in the system (Castor Aa and Castor Ba). The two largest stars in this system are much bigger and brighter than our own sun. This system is located 51 light-years away, and it is best seen in the winter months in the northern hemisphere.

STAR CHARACTERISTICS

Star Name	Castor
Rank	24th brightest star in the night sky
Pronunciation	CASS-ter
Star Name Meaning	Castor, figure in Greek mythology
Alternative Names	Alpha Geminorum
Constellation	Gemini the Twins
Location (Right Ascension, Declination)	$7^h\ 34^m\ 36^s$ +31° 53′ 18″
Distance	Approx. 51 light-years
Magnitude	Apparent: +1.93 — Absolute: +0.59

System	Six star system: Castor Aa—white main sequence star Castor Ab—main sequence red dwarf star Castor Ba—white main sequence star Castor Bb—main sequence red dwarf star Castor Ca—main sequence red dwarf star Castor Cb—main sequence red dwarf star
Temperature★	10,283 Kelvin
Spectral Class★	A-type star
Visible From	Northern Hemisphere: 0° to +90°N / Southern Hemisphere: 0° to -60°S
Compared to Our Sun★	Mass: 2.15 Solar Masses / Luminosity: 34 times brighter than the sun / Radius: 2.4 Solar Radii
Exoplanets	None detected
Eventual Fate★	Red giant, planetary nebula, then white dwarf
Interesting Facts	✦ Pollux is considered the "twin" of Castor. Both stars represent the heads of the twins of the Gemini constellation. Pollux is the 17th brightest star in the night sky, while Castor is the 24th brightest. ✦ While Pollux and Castor are positioned close to each other in the night sky, they are not near each other in space. Castor is approximately 51 light-years away, while Pollux is roughly 33 light-years away. ✦ Despite having the alpha designation of Gemini, it is the second-brightest star in the constellation.

★Primary star Castor Aa

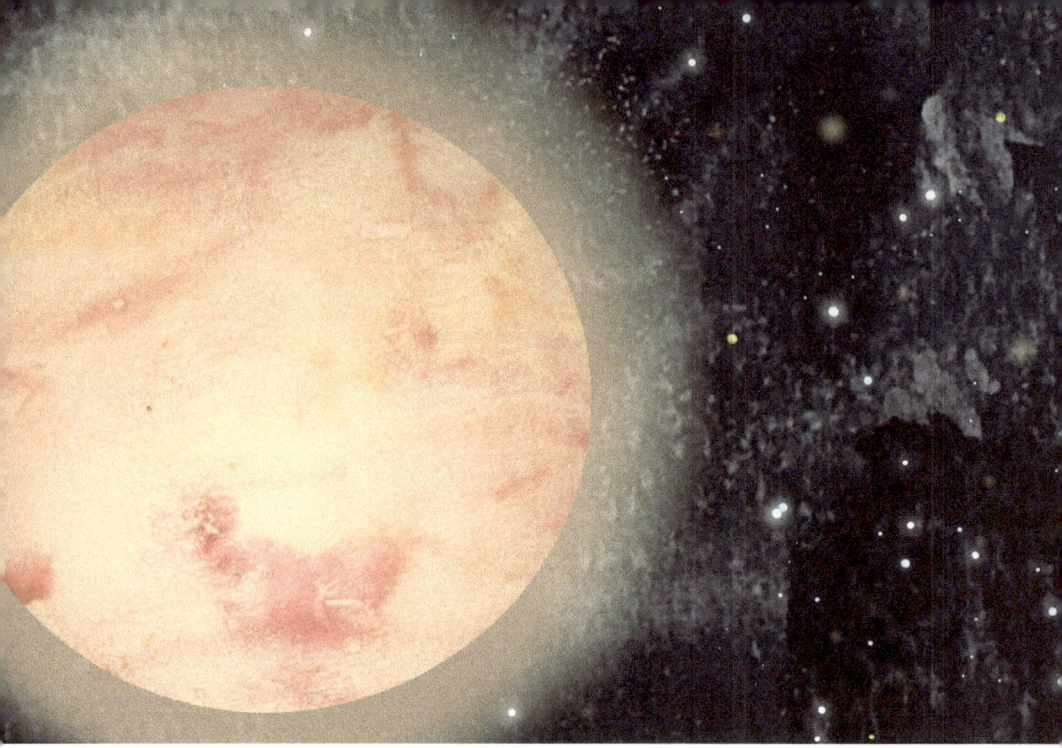

Gacrux (Crux)—The Closest Giant Star

Gacrux, also known as Gamma Crucis, is one of the brightest stars in Crux, the Southern Cross, and is located near the top of the cross-shaped pattern. Gacrux is the third brightest star in the Crux constellation and the twenty-fifth brightest star in the night sky.

At 88 light-years away, it is classified as a red giant star, indicating that it is nearing the end of its life cycle. This star's core has run out of hydrogen and is now fusing helium to produce carbon and oxygen. Following the red giant phase, Gacrux will shed its outer layers and evolve into a planetary nebula, then a white dwarf. The star's remaining core will remain hot until it cools and no longer produces light. Gacrux is an example of what our sun will eventually evolve into billions of years from now.

STAR CHARACTERISTICS

Star Name	Gacrux	
Rank	25th brightest star in the night sky	
Pronunciation	GAK-kruks	
Star Name Meaning	From the Greek words "ga" from gamma and "crux" from the constellation	
Alternative Names	Gamma Crucis	
Constellation	Crux the Southern Cross	
Location (Right Ascension, Declination)	$12^h\ 31^m\ 10^s$ -57° 6' 48"	
Distance	88.6 light-years	
Magnitude	Apparent: +1.63	Absolute: -0.52

25 BRIGHTEST STARS IN THE NIGHT SKY

System	Single star system: Gacrux—orange giant star
Temperature	3,600 Kelvin
Spectral Class	M-type star
Visible From	Northern Hemisphere: 0° to +20°N / Southern Hemisphere: 0° to -90°S
Compared to Our Sun	Mass: 1.5 Solar Masses / Luminosity: 758 times brighter than the sun
	Radius: 120 Solar Radii
Exoplanets	None detected
Eventual Fate	Red giant, planetary nebula, then white dwarf
Interesting Facts	• Third brightest star in the constellation Crux, the Southern Cross. • Nearest red giant to our own sun. • This star was once visible to the Greeks and Romans, but over time this star was forgotten by those living in the European latitudes. The slow precession of the Earth's axis over time meant that the star stopped rising above the horizon in the mid-northern latitudes, and only those living close to the equator or in the southern hemisphere could see it.

Ranking according to distance from our Sun

- Mimosa
- Arcturus
- Fomalhaut
- Rigel Kentaurus
- Gacrux
- Sirius
- Altair
- Aldebaran
- Vega
- Procyon
- Acherr
- Regulus
- Pollux
- Capella
- Castor
- Spica

5 Light Years
15 Light Years
50 Light Years
100 Light Years

Canopus

Betelgeuse

Hadar

Antares

Rigel

Adhara

500 Light Years

Acrux

Deneb

Shaula

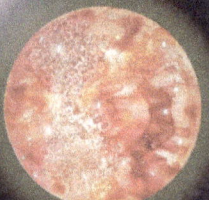

Single Star Systems

Altair
Arcturus
Betelgeuse
Achernar
Canopus
Deneb
Sirius
Aldebaran
Gacrux
Pollux
Vega

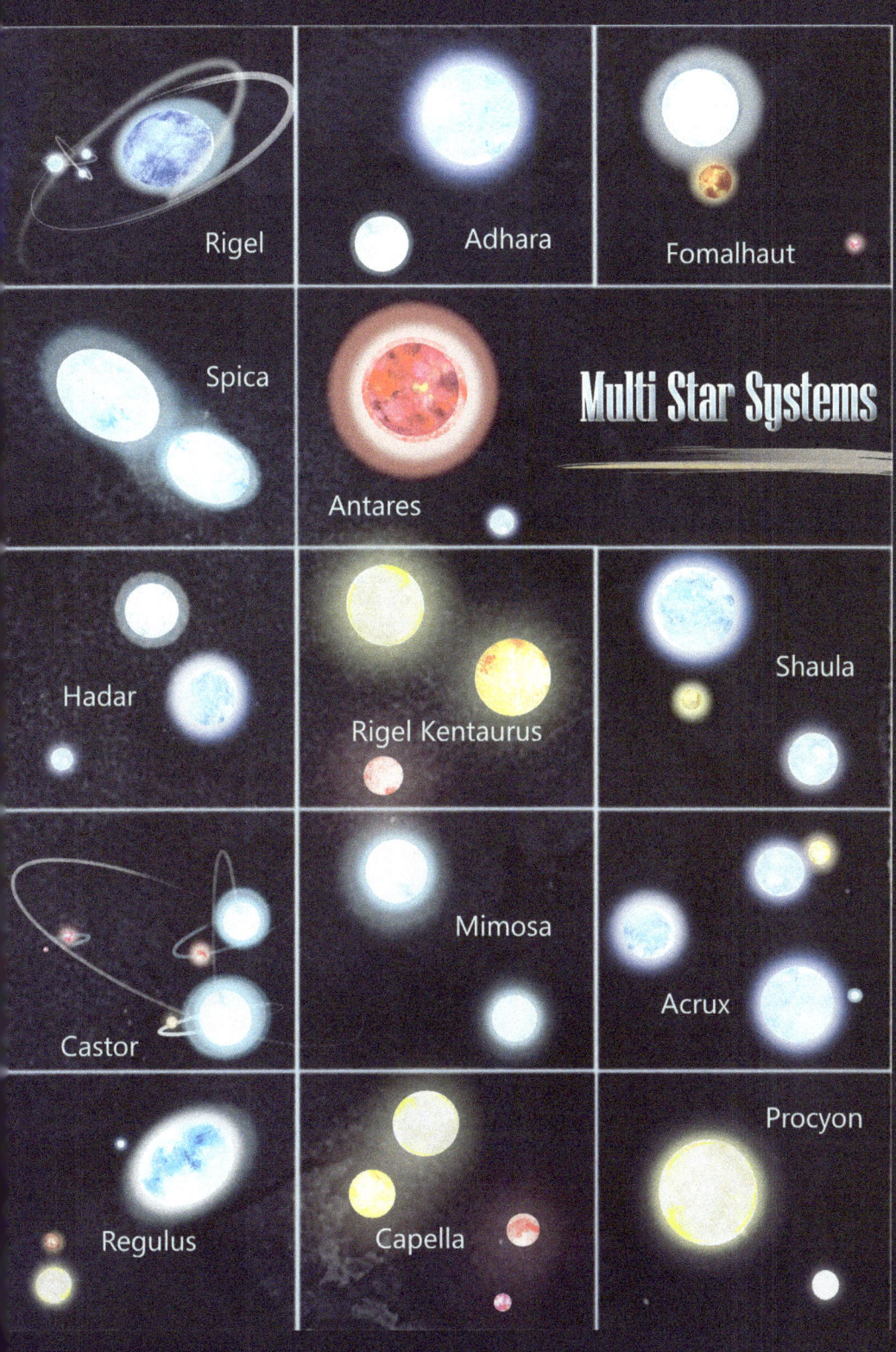

UNIQUE STARS

Proxima Centauri System

Closest Star: Proxima Centauri

Proxima Centauri, also called Alpha Centauri C, is Earth's nearest star system, approximately 4.24 light-years distant. It is categorized as a red dwarf star, which is far smaller than our own sun. Proxima Centauri is part of the Alpha Centauri star system, which also contains Alpha Centauri A and B. This stellar system contains a total of three stars. Proxima Centauri is not visible to the naked eye, however, the main star of this system, Alpha Centauri A, is listed as the third brightest star in the night sky. Proxima Centauri has one-eighth the mass of the sun and shines in a red-hued color.

As of 2022, three exoplanets had been identified around Proxima Centauri:

- **Proxima Centauri b:** This is a potentially Earth-like planet located within the star's habitable zone. It has about the same mass as Earth and orbits the star roughly every eleven days.

This was the first planet to be discovered near Proxima Centauri in 2016. It is unclear at this time if it has an atmosphere because it receives much more radiation than Earth, which could prevent it from having stable living conditions.

- **Proxima Centauri d:** This exoplanet candidate closely orbits Proxima Centauri. It was the last exoplanet to be found in this system, hence the label of "d." It is one-quarter the mass of the Earth, and orbits the star every five Earth days. Since this planet is so close to the red dwarf star, it is not likely to be a habitable planet.

- **Proxima Centauri c:** This exoplanet candidate is further away from the red dwarf star and has a mass of about 7 Earth masses. It is possible this exoplanet is either a Super Earth or a Neptune-like gas giant. It orbits the star every 5.2 Earth years.

Proxima Centauri is a known "flare star," which means that its radiation output is unpredictable, causing its brightness to fluctuate. Scientists believe that the random outbursts of energy are triggered by a star's magnetic field activity.

Smallest: EBLM J0555-57Ab (57 Ab, for short)

This tiny star was discovered by scientists looking for exoplanets (planets outside the solar system). It is located in the modern constellation called Pictor the Easel. Its size is comparable to Saturn; however, its mass is 250 times more than the famous ringed planet. It is the tiniest star known and it is part of a triple star system. It is estimated to be 600 light-years away from Earth. Despite its tiny size in comparison to other stars, the process of hydrogen fusion is able to take place. Scientists originally thought this star was a planet, but after calculating its mass, it was determined that 57Ab was indeed a star.

RMC 136a1: The Most Massive Star Yet Known

The most massive, brightest, and hottest star known is RMC 136a1 and is located in the constellation Dorado the swordfish. This star is thought to be over 320 solar masses and 4.7 million times brighter than our sun. RMC 136a1, also known as R136a1 for short, has an estimated temperature of 46,000 Kelvin and is located 163,000 light-years away from Earth. R136a1 is also part of NGC 2070, an open star cluster in the Tarantula Nebula. It is the biggest of the two hundred stars in the cluster.

R136a1 is classed as a Wolf-Rayet star, a rare type of star that contains heavier elements like carbon and nitrogen. The most massive stars have very brief lives, ending explosively as a supernova and finally collapsing into a black hole. Giant stars live for millions of years, but ordinary stars, such as our sun, survive for billions of years.

When you zoom in from the Tarantula Nebula to the R136 cluster, you can see R136a1/2/3 as the slightly resolved area in the bottom right corner.

It is likely that by the time this book is published, there will be a new candidate for the "most massive star known." Astronomy is an ever-changing field with many new discoveries. As technology continues to advance so will our understanding of the unique stars of our universe.

Mira: A Wonderful Star with a Tail

Mira, commonly known as Omicron Ceti, is a peculiar star in the Cetus constellation. It is a binary star system consisting of Mira A, a red giant pulsating star, and Mira B, a white dwarf star orbiting the primary star Mira A. Astronomers have the nickname Mira for this star, which is a Latin name meaning "wonderful."

Mira A is in the final stages of its life as a red giant star. It has depleted the hydrogen fuel in its core and has grown into a massive, cooler star. Mira's surface temperature is low when compared to other stars, giving it a reddish color. Mira is also shedding its outer layers into space, forming a shell of gas called the circumstellar envelope. But what is unique about this star is that it appears to have a long jet of material that is being ejected away from the dying star.

This "tail" is a stream of material trailing behind the star as it moves through space. The tail of Mira A is measured to be 13 light-years long, which is approximately 20,000 times the average distance from the sun to Pluto. It consists of streams of gas and dust that have been ejected from Mira over thousands of years. Mira A is also traveling at a fast speed of 219,000 kilometers per hour, which is faster than a speeding bullet!

Mira, commonly known as Omicron Ceti

Lucy (in the Sky with Diamonds)

BPM 37093 (also called V855 Centauri) is a white dwarf star, which is the leftover core of a star that has depleted its nuclear fuel and lost its outer layers. The white dwarf stage of a star is the last stage of its life and visibility. Once a white dwarf cools down it will become invisible. This star is located in the constellation of Centaurus and is estimated to be 48 light-years away from Earth.

BPM 37093 is particularly notable for its crystalline core. The core is made up of carbon and oxygen, which have hardened into a crystalline structure, resembling a gigantic diamond. BPM 37093 is not visible to the naked eye but can be studied with the aid of telescopes equipped with instruments capable of detecting its faint light and subtle variations.

This star offers the first direct proof that white dwarfs crystallize as they cool. It has been informally dubbed "Lucy" after the Beatles' song "Lucy in the Sky with Diamonds." Astronomers can explore the internal structure of white dwarfs by studying BPM 37093's pulsations using a technique known as asteroseismology. Scientists can learn about the internal density, temperature, and composition of a star by analyzing its pulsation modes.

Scientists estimate that 90 percent of the core of BPM 37093 is crystalline in structure. This makes it one of the most intriguing and "valuable" objects in terms of material composition in the universe.

Algol: The Demon Star

Algol, also nicknamed the Demon Star, is the second-brightest star in the constellation of Perseus. Algol is an excellent example of a variable star since its brightness can change throughout the night. Algol is associated with the Greek myth of Perseus and the Gorgon Medusa. In this story, Perseus beheads Medusa, a snake-haired creature whose gaze could turn people to stone. Algol represents the eye or head of Medusa, which Perseus is said to hold in his hand. Algol is often referred to as the "Demon Star" due to its unpredictable brightness changes and its association with myths and folklore.

Algol is a three-star system composed of the following components:

1. Algol A: The primary component, a bright blue-white main sequence star, about 3.5 times more massive and about 180 times more luminous than the sun.

2. Algol B: The secondary star, an orange subgiant, less massive and cooler than Algol A, but much larger in size.

3. Algol C: A more distant companion star that is also a main sequence star, orbiting the central pair.

The Algol system is approximately 90 light-years away from Earth and has an apparent magnitude that varies between 2.1 and 3.4, making it visible to the naked eye. The stars Algol A and Algol B are eclipsing binary stars that orbit each other in such a way that they periodically pass in front of each other from our point of view on Earth. When the dimmer Algol B passes in front of the brighter Algol A, the system's overall brightness decreases significantly, causing the observed "eclipse" and a drop in apparent magnitude.

In many cultures, Algol is connected with stories of demons, spirits, or monsters. In Greek mythology, Algol is often represented as the head of Medusa. In Hebrew folklore, Algol was called Rōsh ha Sāṭān, or "Satan's Head." The name Algol is derived from the Arabic term ra's al-ghul, which means "head of the ogre."

NOTABLE STAR CLUSTERS

Pleiades

The Pleiades Star cluster is one that has captured the attention and imagination of humans for thousands of years. It is one of the few celestial objects that can be seen with the unaided eye. It looks like a small, cloudy patch in the night sky and under dark skies, several individual stars can be distinguished. From the vantage point of Earth, it looks like the cluster has between six to seven stars, but in reality it has over one thousand stars! The "Seven Sisters" star cluster is estimated to be approximately 440 light-years from Earth. The Pleiades are visible starting in November into March (this can vary depending upon your latitude).

The main stars of the system are classified as B-type blue giant stars. These types of stars live quickly and die explosively. The cluster is also surrounded by nebulous gas and dust and light from the stars reflects on the gas causing it to glow. It is predicted that this star cluster will eventually disperse as all star clusters do over time.

The earliest record of the Pleiades comes from the Caves of Lascaux in Montignac, France. This site contains prehistoric cave paintings that date back to over 17,000 years ago, placing it in the Upper Paleolithic time period. This expansive system of caves features many animals, including horses, deer, aurochs (extinct wild cattle), bison, and ibexes. There are also depictions of mythical creatures, humans, and abstract symbols. The Pleiades appear over a painting of a bull. This cave system was discovered in 1940, but was ultimately closed off to the public more than twenty years after its discovery in order to protect the paintings from carbon dioxide, molds, and bacteria.

The paintings in the Lascaux Caves in France depict a bull alongside a cluster of dots, which is believed to represent the Pleiades.

Another early representation of the Pleiades is the Nebra Sky Disk. This ancient artifact is a bronze disk around thirty centimeters (twelve inches) in diameter with a blue-green patina and gold symbols inlaid into the medallion. It was discovered in 1999 near Nebra, Germany, and dates from around 1600 BCE, placing it in the Bronze Age.

This relic is regarded as one of the oldest known representations of the cosmos.

The Pleiades is represented by the grouping of seven stars between the sun (full circle shape) and the moon (crescent shape). The actual purpose of the Nebra Sky Disk is still debated among experts, but it is widely assumed to be an astronomy instrument or a ritual device.

Many cultures have their own names and stories for this famous star cluster. In Japan, the Pleiades are known as "Subaru," which means "gathered together."

The Māori people of New Zealand call the Pleiades Matariki. In New Zealand, this cluster can be seen all year long except for about a month in the winter. To many Māori, the first time they see the cluster in the sky, in late June or early July, is the start of the Māori new year, which is also known as Matariki.

The Hawaiians called the Pleiades Makali'i, and it signals the start of the Hawaiian New Year and as well as new weather patterns and harvesting island crops.

The Aztec and Mayan people in Middle America built their calendars around the movement of the Pleiades. Every fifty-two years, the constellation would pass exactly over important sites for ceremonies in the Mayan culture. The Pleiades can also be seen in the temple complex of Teotihuacan, which is close to Mexico City.

In South Africa, the Pleiades were known as IsiLimela, or the "digging stars," because their presence signaled the need to start hoeing the ground. These stars were used to signal the start of the agricultural season throughout Africa.

Cluster Characteristics	
Name(s)	Pleiades, Messier 45, Matariki, Makali'i, Subaru, Seven Sisters
Type	Open star cluster
Distance	444 light-years
Constellation	Taurus
Visibility	November to March
Magnitude	+1.6

Hyades

The Hyades star cluster is among the most studied open star clusters because of its age and close vicinity to Earth. It is located in the constellation of Taurus and lies in close proximity to the Pleiades Star cluster. These two clusters make up the asterism known as the "Golden Gate of the Ecliptic" because the sun, moon and planets all pass through these two celestial objects. Aldebaran, the bright red giant star in Taurus and the 14th brightest in the night sky, appears to be part of the cluster, but it is actually a foreground star, much closer to

Earth than the cluster itself. This cluster is easy to point out in the sky because it makes a distinctive V-shape, making up the face of Taurus the Bull, while Aldebaran represents the red eye of the bull.

Scientists have estimated the age of the cluster to be around 625 million years old and it contains hundreds of stars, including main-sequence and red giant stars. Several dozen stars are visible to the naked eye under good conditions; however, this cluster is best viewed with binoculars. The Hyades has a dense core of about 10 light-years in diameter and a more extended halo reaching up to 60 light-years across.

The bright star Aldebaran with the Hyades star cluster is seen on the left, and the Pleiades to the right in the constellation Taurus. The red planet Mars moves between these two clusters in the sky. Together the Pleiades and Hyades make up the asterism known as the "Golden Gate of the Ecliptic."

Cluster Characteristics	
Name(s)	Hyades
Type	Open
Distance	153 light-years away
Constellation	Taurus
Best Visibility	November to March
Magnitude	+0.5

Beehive Cluster

The Beehive Cluster, also known as Praesepe or Messier 44, is one of the nearest and best-known open star clusters to Earth. It is located in the constellation of Cancer, and it is estimated to be 600 light-years from Earth. The name "Praesepe" means "manager" in Latin, and "Beehive Cluster" comes from its appearance resembling a swarm of bees. The Beehive Cluster contains over a thousand stars, with around two hundred visible through small telescopes. It includes a mix of main sequence stars, red giants, and white dwarfs stars.

Cluster Characteristics	
Name(s)	Beehive Cluster, Messier 44, Praesepe
Type	Open cluster
Distance	600 light-years
Constellation	Cancer
Best Visibility	March, April, May in northern hemisphere
Magnitude	+3.7

Coma Cluster

The Coma Star Cluster, also known as the Coma Berenices Cluster or Melotte 111, is an open star cluster located in the constellation Coma Berenices. It is one of the nearest open clusters to Earth at a distance of 280 light-years. This star cluster was once represented as the tail of Leo the Lion, but in modern times it is part of Coma Berenices. The Coma Star Cluster contains about forty to fifty stars that are visible to the naked eye, and many more that can be observed with binoculars or a telescope. The stars are primarily main sequence stars, but there are also some giant stars scattered throughout this grouping.

The Coma Cluster should not be confused with the Coma Berenices Galaxy Cluster, which are often referred to by the same name. The Coma Galaxy Cluster is a vast grouping of distant galaxies. More than a thousand galaxies have been confirmed and discovered in this galactic cluster.

The Coma Star Cluster is the grouping of stars toward the left upper side of the photo. It is visible to the unaided eye in areas with dark skies.

The Coma Galaxy Cluster is a grouping of more than a thousand galaxies. Ancient observers were not able to see these galaxies. Modern astronomers use technology and special analysis techniques in order to examine what the eye cannot see.

Cluster Characteristics	
Name(s)	Melotte 111
Type	Open star cluster
Distance	280 light-years away
Constellation	Coma Berenices
Best Visibility	March to June
Magnitude	+1.8

Alpha Persei Cluster

The Alpha Persei Cluster, also known as Melotte 20 or Collinder 39, is an open star cluster located in the constellation Perseus. It is one of the more prominent clusters visible to the naked eye. The cluster contains several bright stars, including many B-type main sequence stars, with the brightest member being Alpha Persei, also known as Mirfak. It is visible to the naked eye, however, looking at it with binoculars or small telescopes will reveal more stars within the cluster.

Cluster Characteristics	
Name(s)	Melotte 20
Type	Open star cluster
Distance	570 light-years away
Constellation	Perseus
Best Visibility	November to March
Magnitude	+1.2

Hercules Cluster

The Hercules Star Cluster, also known as the Hercules Globular Cluster or M13 (Messier 13), is one of the most visible and most researched globular clusters in the northern hemisphere. Messier 13 is made up of many hundred thousand stars that are densely packed together to form a sphere. It consists of a variety of stars, with the majority being ancient, low-metal stars. The cluster is approximately 145 light-years in diameter, with a core radius of 1.7 light-years. The cluster has an apparent magnitude of 5.8, making it visible to the human eye, but only in very dark skies. It is best seen using binoculars or a telescope, which highlight its dense, star-filled core. In the northern hemisphere, the Hercules Cluster is most visible in late spring and summer. In 1974, the Arecibo Observatory utilized M13 as the target for the Arecibo Message, a radio message meant for potential extraterrestrial civilizations.

Cluster Characteristics	
Name(s)	Messier 13, Great Hercules Cluster
Type	Globular Cluster
Distance	25,000 light-years away
Constellation	Hercules
Best Visibility	March to June
Magnitude	+5.8

Southern Pleiades

The Southern Pleiades, sometimes referred to as the "Theta Carinae Cluster" or IC 2602, is an open star cluster in the constellation Carina the Keel. This star cluster is very similar to the more well-known Pleiades cluster in the northern hemisphere, it is commonly referred to as the Southern Pleiades. Its distance is estimated to be 479 light-years away and contains about seventy-five stars.

This cluster is one that can be seen with the unaided eye and ranks as the fifth brightest star cluster in the night sky. The brightest star in this cluster appears to be Theta Carinae, however this star is not a part of the cluster. Much like the Northern Pleiades, this star cluster is best viewed through binoculars and can only be viewed from the southern hemisphere.

Cluster Characteristics	
Name(s)	IC 2602
Type	Open Star Cluster
Distance	479 light-years
Constellation	Carina
Best Visibility	March
Magnitude	+1.9

FAQ

How many stars are visible in the sky?

If you are stargazing in a very dark place, it is possible to see between five and six thousand stars throughout the night. People with exceptional eyesight may see up to eight thousand stars.

What percentage of stars are in a double or multi-star system?

Scientists estimate that approximately 85 percent of all the stars in the universe are part of a double system or multi-star system.

How do astronomers study light?

Astronomers use a scientific tool called a spectrometer to break light up into its component parts. A star's spectrum gives researchers lots of information about celestial objects. A star's spectrum can reveal how hot it is, how fast it is moving in the direction of travel, and determine the strength of its magnetic field (plus much, much more). Astronomers will also compare a star's spectrum to one created in a lab. This allows scientists to determine what elements are present inside a star.

The Science of Spectroscopy

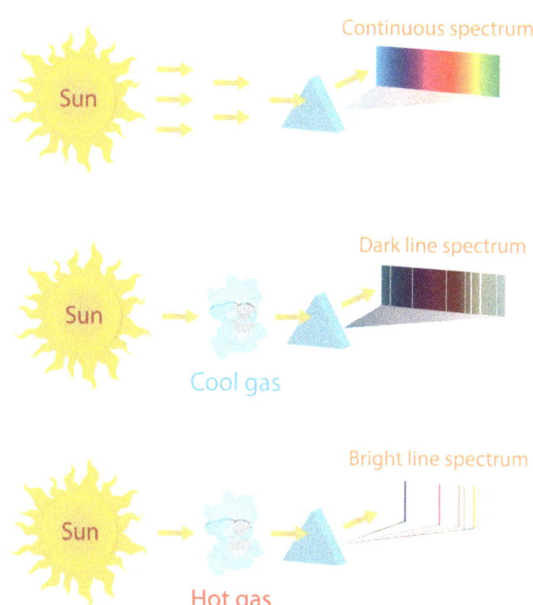

Why are there no green stars?

Stars emit light through a process known as blackbody radiation. The color of this emitted light depends on the star's temperature. Hotter stars emit shorter-wavelength, bluer light, while cooler stars emit longer-wavelength, redder light.

The green color falls in a relatively narrow range between the cooler, redder stars (K and M types) and the hotter, bluer stars (A and B types). Because of the way blackbody radiation works, there are very few stars that have temperatures in the range where they would emit a significant amount of green light. Most stars fall on the red side or blue side of this "green gap."

While there are some stars that may emit a small amount of green light, it is typically overwhelmed by the other colors they emit. Therefore, stars in the night sky generally appear white, yellow, orange, red, or blue, and this ultimately depends on their temperature and spectral type. The lack of green stars in the night sky is a consequence of the way stars' temperatures and colors are distributed across the electromagnetic spectrum.

What is light pollution?

Light pollution is artificial nighttime lighting in urban and suburban areas that interferes with sky observation. Light pollution is highest in metropolitan areas. When trying to observe the stars, planets, and celestial objects, it is best to seek out dark-sky areas and avoid light pollution. Stargazing is at its best away from light pollution. This includes both natural and artificial light. Stargaze when the moon is absent, and travel away from city lights to get the best view.

What creates the glow of the Milky Way?

When we look at the Milky Way, we are observing the combined light of billions of stars in our galaxy's plane. These stars are too far away for our eyes to resolve single points of light. All the stars of the Milky Way create a glow that is best seen in dark-sky areas.

How is astrology different from astronomy?

Astrology is the study that attempts to interpret the influence of the heavenly bodies on human affairs. Ancient humans studied the sky to make sense of the world around them and used it as a calendar to mark the changing seasons and to time agricultural activities.

In ancient times, astrology and astronomy were considered the same branch of study. Humans also connected stories and legends to the star patterns. These stories could range from morality tales to love stories and often predict the changing seasons. These stories were passed down from generation to generation, thus preserving important cultural knowledge.

However, since the Enlightenment period of the seventeenth century, astrology and astronomy have diverged. As observational technology improved and with the development of the scientific method, many of the ideas behind astrology began to be debunked.

In modern times, astronomy is the study of the material world beyond the Earth's boundaries. There are many branches of astronomy, including sky observation and research, human spaceflight, and space technology. Each sub-topic of astronomy works together to reach a common goal—to learn more about the world beyond our own.

It is important to remember that astrology and astronomy are no longer the same branch of study as they used to be centuries ago. Many astronomers consider astrology to be a pseudoscience. But even though these branches of study are different, they are connected in origin.

Is it possible to observe your zodiac constellation on your birthday?

No, you cannot see your zodiac constellation on your birthday because the sun is traveling through it at that time. Since the sun is "in" the constellation at the time of your birthday, the star pattern is out during the daytime and therefore cannot be seen. Astronomers classify constellations based on their position in the sky. If a constellation intersects with the path of the sun, it is classified as a zodiacal constellation. The optimal time to view your constellation is typically four to six months after your birthday, though this may vary based on your latitude, geographic location, constellation size, and its duration of visibility in the night sky.

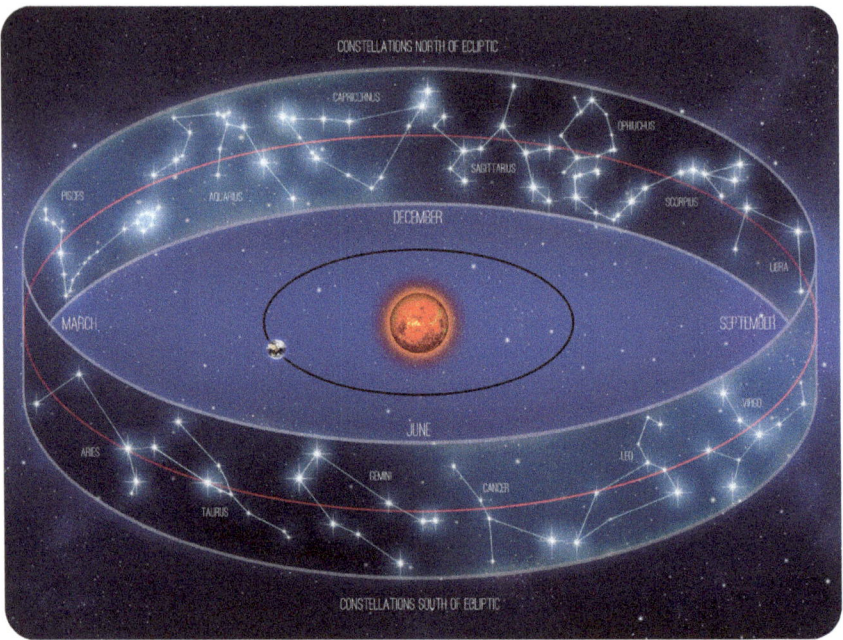

Sometimes a person's star sign does not line up with what is actually happening in the nighttime sky. The discrepancy between zodiac signs and their corresponding constellations arises due to a phenomenon called precession of the equinoxes and the historical way the zodiac was defined. The zodiac is a band of twelve signs along the ecliptic (the apparent path the sun takes across the sky over the course of a year), divided into equal segments of 30° each. This division was established by the Babylonians around three thousand years ago. When the Babylonians established the zodiac, they divided the ecliptic into twelve equal segments based on the sun's position at the time of the vernal equinox. At that time, the constellations roughly aligned with these twelve segments. However, constellations are not uniform in size or shape, and there are actually thirteen constellations along the ecliptic path, not just twelve. The constellation Ophiuchus, for example, lies along the ecliptic but is not included in the traditional zodiac. Due to precession of the equinoxes, the position of the sun relative to the background stars changes over time. As a result, the sun now appears in a different part of the sky compared to where it was about two thousand years ago, when the zodiac was first established. This means that the sun currently appears in different constellations than those traditionally associated with each zodiac sign. Over the centuries, the cumulative effect of precession has shifted the zodiac signs by about one month in relation to the constellations.

Why do stars twinkle?

Stars appear to twinkle because they are points of light, unlike planets that are disk-like in shape in the sky. The light of the stars passes through our atmosphere, and the air can behave like a moving lens, causing the stars to look like they are twinkling.

How do stars produce their energy?

A star produces its energy through the process of nuclear fusion. In the center core of the star, extreme pressure causes hydrogen atoms to collide at high speeds. When 4 protons fuse, helium is generated and energy is produced. This allows a star to produce light and other forms of electromagnetic energy.

How are planets different from stars?

Planets do not generate their own light. Some planets, like Jupiter, can radiate infrared and radio waves. Planets are also much smaller in size and contain less mass than stars. From a stargazing perspective, planets do not appear regularly each season the way the stars do. Planets are always in motion, orbiting the sun. In the sky, planets always move along the plane of the sky known as the ecliptic.

Why is it difficult to determine the distance of some stars?

Determining the distance between stars is challenging due to a variety of factors, including measurement difficulties, the vastness of space, and the nature of the stars themselves. Distant stars are often faint, making it difficult to obtain accurate measurements of their properties. Observing faint stars requires long exposure times and sensitive detectors. In regions of the sky with many stars (e.g., the galactic plane), distinguishing individual stars and accurately measuring their properties can be challenging due to overlapping light sources.

Will the sun explode into a supernova?

No, the sun will not explode into a supernova. The sun is not massive enough to undergo the supernova process. Supernovae typically occur in stars with initial masses greater than eight to ten times the mass of the sun. These massive stars end their lives in a supernova explosion when they exhaust their nuclear fuel and undergo gravitational collapse. The sun, with a mass of approximately 1 solar mass, is well below this threshold.

APPENDIX

EQUATORIAL STAR MAP

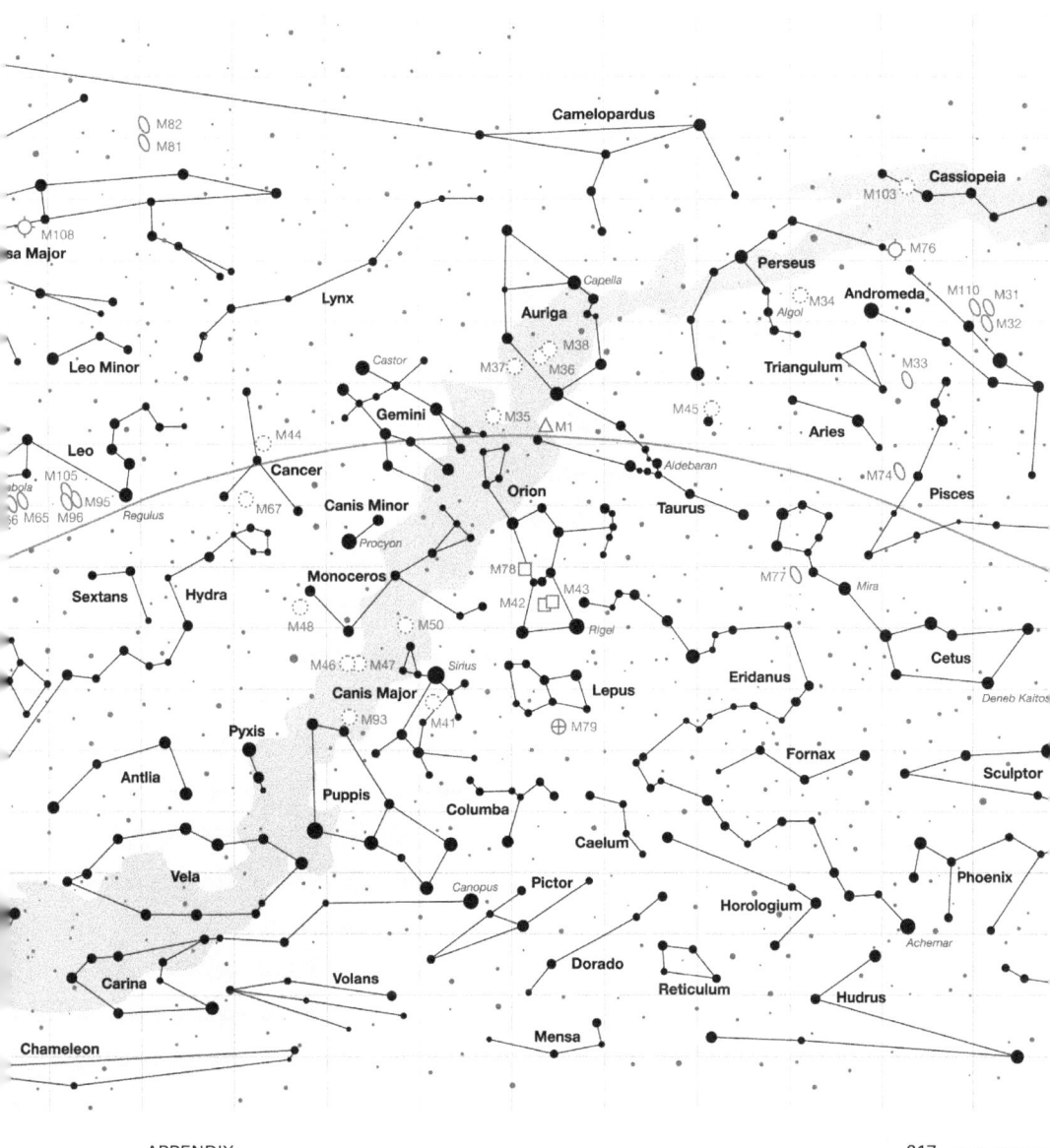

CONSTELLATION PICTURE REFERENCE GUIDE

Aquila
Star: Altair

Auriga
Star: Capella

Boötes
Star: Arcturus

Canis Major
Stars: Sirius, Adhara

Canis Minor
Star: Procyon

Carina
Star: Canopus

Centaurus
Stars: Rigil Kentaurus, Hadar

Crux
Stars: Acrux, Mimosa, Gacrux

Cygnus
Star: Deneb

Eridanus
Star: Achernar

Gemini
Stars: Pollux, Castor

Leo
Star: Regulus

Lyra
Star: Vega

Orion
Stars: Rigel, Betelguese

Piscis Austrinus
Star: Fomalhaut

Scorpius
Stars: Antares, Shaula

Taurus
Star: Aldebaran

Virgo
Star: Spica

GLOSSARY

+ **Absolute magnitude:** Absolute magnitude is the intrinsic brightness of a celestial object, such as a star, measured as if it were viewed from a standard distance of 10 parsecs.

+ **Apparent magnitude:** Apparent magnitude is a measure of how bright a celestial body appears to an observer. For historical reasons, the magnitude scale assigns larger numbers to fainter objects, making it an inverse scale.

+ **Asterism:** Asterisms are patterns of stars that can be seen in the night sky. They are collections of stars and patterns that are not recognized as part of the eighty-eight official constellations accepted by the International Astronomical Union (IAU). Asterisms can be a part of a constellation, made up of stars from different constellations, or form recognizable patterns and shapes.

+ **Astronomer:** An astronomer studies the objects that make up the universe. Astronomers can be professionals or amateurs.

+ **Astronomy:** Astronomy is the scientific study of celestial objects including but not limited to, stars, planets, moons, and galaxies. It involves observing and understanding how celestial objects form, move, and interact.

+ **Astronomical Unit:** An Astronomical Unit (AU) is a standard measurement used in astronomy to describe distances within our solar system. It is defined as the average distance from the Earth to the sun, which is about 93 million miles (or 150 million kilometers). Astronomical units are used to help scientists compare distances between celestial bodies.

- **Asteroseismology:** Asteroseismology is the study of the internal structure and processes of stars through their oscillations and vibrations. Scientists can learn more about a star's composition, age, and other characteristics by observing how stars pulse and change brightness over time.

- **Aurora:** An aurora is a natural light display predominantly seen in high-latitude regions around the Arctic and Antarctic. These displays are caused by the interaction of charged particles from the solar wind with the Earth's magnetic field and atmosphere.

- **Black holes:** A black hole is an object with a gravitational pull so strong that nothing, not even light, can escape from it. Most black holes form from the remnants of a massive star that has ended its life cycle. After exhausting its nuclear fuel, the star undergoes a supernova explosion, and if the remaining core is sufficiently massive, it collapses under its own gravity to form a black hole.

- **Cepheid variable stars:** A type of pulsating star known for their predictable brightness variations.

- **Constellation:** A constellation is a group of stars that forms a recognizable pattern in the night sky and are typically associated with mythological figures, animals, or objects. Some are ancient in origin while others are modern. Constellations are used for storytelling, navigation, and calendar systems.

- **Ecliptic:** Apparent path of the sun, moon and planets in the sky.

- **Electromagnetic radiation:** Electromagnetic radiation is a type of energy that propagates over space as oscillating electric and magnetic fields. The electromagnetic spectrum consists of a wide range of wavelengths and frequencies.

- **Electron:** An electron is a small, negatively charged particle that orbits the nucleus of an atom.

- **Exoplanet:** A planet located outside the local solar system.

- **Fusion:** Nuclear fusion is the process where the atomic nuclei of lighter elements join together to form the nucleus of a heavier element.

- **Galaxy:** A galaxy is a system of stars and other material components such as dark matter, gas, and dust that are gravitationally bound, and usually separated from its neighbors by hundreds of thousands of light-years.

- **Gnomon:** Sticks that produce shadows and are used for telling time.

- **Gravity:** Gravity is the mutual attraction of objects with mass. It is an invisible force the extends infinitely in the universe.

- **Helioseismology:** Helioseismology is the study of the internal structure and dynamics of the sun through the observation of its oscillations or "solar waves."

- **Helium:** Helium is the second most abundant element in the universe, following hydrogen. Helium is produced in the core of stars through hydrogen fusion.

- **Hydrogen:** Hydrogen is the most abundant element in the universe. Hydrogen is the primary fuel for stars. In the cores of stars, hydrogen nuclei (protons) fuse to form helium through nuclear fusion, releasing immense energy.

- **Magnitude:** Measure of a star's brightness. There are two main types of magnitudes used: apparent magnitude and absolute magnitude.

- **Main sequence stars:** A main sequence star is a star that is in the longest and most stable phase of its life cycle, where it is fusing hydrogen into helium in its core. The forces of fusion and gravity are balance, making the star stable.

- **Meteor:** A meteor is a small piece of rock or metal from space that enters into Earth's atmosphere, producing a bright streak of light as it burns up due to friction with the air.

- **Meteoroid:** A meteoroid is a small rock or metallic body in space that is smaller than an asteroid. The size of a meteor typically ranges from a grain of dust to a few meters across. Meteoroids come from comets, asteroids, or other celestial objects. When a meteoroid enters the Earth's atmosphere it is then considered a meteor.

- **Meteorite:** A meteorite is any part of a meteor that lands on Earth's surface after traveling through Earths' atmosphere.

- **Moon:** The moon is Earth's only natural satellite and is the fifth-largest moon in the solar system. It plays a significant role in various aspects of Earth's natural processes and has been a subject of human fascination and exploration for centuries.

- **Nadir:** In astronomy, nadir refers to the direction that is straight down from where a person is standing. It is the lowest point in the sky relative to an observer.

- **Neutron:** A neutron is a small, neutral particle found in the nucleus of an atom. Neutrons significantly contribute to the total mass of an atom.

- **Neutron stars:** A neutron star is a compact, dense remnant of a massive star that has undergone a supernova explosion.

- **Occultation:** An occultation occurs when one celestial body moves in front of another, temporarily blocking the view of the latter from a specific point of observation. This can occur with stars, planets, the moon, and other celestial objects.

- **Planet:** A planet is a celestial body that orbits a star and has sufficient mass to assume a nearly round shape due to its own gravity.

- **Plasma:** In astronomy, plasma is a state of matter where gasses become ionized. When this happens atoms lose some of their electrons and become charged particles. Plasma is often found in the stars, where extreme temperature and pressure causes this ionization.

- **Precession:** In astronomy, precession refers to the gradual shift in the orientation of an astronomical body's axis of rotation. This affects the way stars appear in the sky over long periods of time.

- **Proton:** A proton is a small, positively charged particle found in the nucleus of an atom. A proton significantly contributes to the weight of an atom. The amount of protons in an atom is used to determine an atom's identity.

- **Protostar:** An early stage in the formation of a star, occurring after a molecular cloud collapses but before nuclear fusion begins in its core.

- **Pulsars:** Pulsars are highly magnetized, rotating neutron stars that emit beams of electromagnetic radiation out of their magnetic poles. These beams are detectable when they sweep past Earth, much like the beam of a lighthouse.

- **Red dwarf star:** Red dwarf stars are the smallest and coolest type of main sequence stars, classified as spectral type M.

- **Red giant stars:** Red giant stars are a late stage in the evolution of stars that have exhausted the hydrogen fuel in their cores.

- **Revolution:** The movement of celestial objects around other objects due to gravitational forces.

- **Rotation:** The spinning motion of a celestial object around its own axis.

- **Satellite:** An object that orbits another object due to gravitational attraction. Satellites can be natural, like moons, or artificial, created by humans for various purposes.

- **Spectral classification:** Spectral classification is the process of categorizing stars based on their light spectra, which are patterns of colors and wavelengths that they emit. Spectra are used to determine a star's temperature, composition, and other characteristics.

- **Spectroscopy:** A scientific technique used to study the interaction between light and matter.

- **Star cluster:** A star cluster is a group of stars that are gravitationally bound and formed from the same molecular cloud.

- **Star:** A celestial body undergoing fusion.

- **Supergiant star:** An exceptionally large and luminous star, much larger and brighter than our sun. This star type has evolved off the main sequence stage and is fusing larger elements in its outer layers. Eventually this type of star will explode into a supernova.

- **T Tauri stars:** T Tauri stars are a class of very young stars that are in the early stages of stellar evolution. They are named after their prototype, T Tauri, located in the constellation Taurus. It is estimated these stars are less than ten million years old.

- **Universe:** The universe is everything that exists, including space, stars, planets, and galaxies. It has no known edge and is always expanding.

- **Variable star:** A variable star is one whose brightness as observed from Earth changes over time. These fluctuations can range from seconds to years and are driven by a variety of phenomena.

- **White dwarf:** A white dwarf is a stellar remnant formed when a low- to intermediate-mass star (such as our sun) depletes its nuclear fuel and expels its outer layers, leaving a hot, dense core.

- **Year:** The time it takes the Earth to complete one trip around the sun.

- **Zenith:** The zenith is defined as the point directly above the observer.

REFERENCES

"Achernar (α Eridani) | Facts, Information, History & Definition." *The Nine Planets*, 27 January 2020, nineplanets.org/achernar-%CE%B1-eridani/. Accessed 19 October 2023.

BBC News. "BBC NEWS | Science/Nature | Diamond star thrills astronomers." *Home—BBC News*, 16 February 2004, news.bbc.co.uk/2/hi/3492919.stm. Accessed 9 December 2023.

Blane, Dave. "Alpha Crucis—Double star of the month." *Astronomical Society of Southern Africa*, 24 February 2014, assa.saao.ac.za/sections/double-and-variable-stars/double-stars/news-and-articles/alpha-crucis-double-star-of-the-month/. Accessed 18 July 2024.

California Institute of Technology Galaxy Evolution Explorer. "Speeding Bullet Star Leaves Enormous Streak Across Sky." *GALEX*, California Institute of Technology, 27 November 2007, www.galex.caltech.edu/newsroom/glx2007-04r.html. Accessed 23 July 2023.

"Castor, the Six-Star System." *Castor, the 6-star System—A3 poster*, Caltech JPL, 10 August 2012, www.jpl.nasa.gov/infographics/castor-the-6-star-system-a3-poster. Accessed 19 July 2023.

Futurism. "Lucy's in the Sky with Diamonds: Meet the Most Expensive Star Ever Found." *Futurism*, 12 June 2014, futurism.com/lucy-in-the-sky-with-diamonds. Accessed 1 March 2024.

Howell, Elizabeth. "Adhara: Brightest Star in Ultraviolet Light." *Space.com*, 27 September 2013, www.space.com/22980-adhara.html. Accessed 7 June 2024.

International Astronomical Union. "Naming Stars | IAU." *International Astronomical Union*, www.iau.org/public/themes/naming_stars/. Accessed 15 April 2024.

"Introduction to Binary Stars." *Australia Telescope National Facility |*, www.atnf.csiro.au/outreach/education/senior/astrophysics/binary_intro.html. Accessed 14 October 2023.

Johnson, Daniel. "Meet Arcturus: Guardian of the Bear—Sky & Telescope." *Sky & Telescope*, 13 October 2020, skyandtelescope.org/astronomy-news/meet-arcturus-guardian/. Accessed 5 September 2023.

Johnson, Daniel. "Meet Canopus, the Second Brightest Star—Sky & Telescope." *Sky & Telescope*, 11 July 2019, skyandtelescope.org/astronomy-news/meet-canopus-second-brightest-star/. Accessed 14 October 2023.

Johnson, Daniel. "Meet Procyon, Orion's Littler Dog—Sky & Telescope." *Sky & Telescope*, 23 March 2022, skyandtelescope.org/astronomy-news/meet-procyon-orions-lesser-dog/. Accessed 18 October 2023.

Johnson, Daniel. "Meet Proxima Centauri: The Closest Star—Sky & Telescope." *Sky & Telescope*, 30 November 2021, skyandtelescope.org/astronomy-news/meet-proxima-centauri-closest-star/. Accessed 5 September 2023.

Johnson, Daniel. "Meet Sirius, the Brightest Star—Sky & Telescope." *Sky & Telescope*, 27 March 2018, skyandtelescope.org/astronomy-news/meet-sirius-brightest-star/. Accessed 19 July 2023.

Kiefert, Nicole. "Astronomers just discovered the smallest star ever." *Astronomy Magazine*, 12 July 2017, www.astronomy.com/science/astronomers-just-discovered-the-smallest-star-ever/. Accessed 12 May 2024.

Kyselka, Will, and Ray E. Lanterman. *North Star to Southern Cross*. University Press of Hawaii, 1976.

Lagaso, Nadine. "The rise of Makaliʻi marks a Hawaiian new year." *Kamehameha Schools*, 19 November 2014, www.ksbe.edu/article/the-rise-of-makalii-marks-a-hawaiian-new-year. Accessed 2 February 2024.

"Meet Spica, the Ear of Grain—Sky & Telescope." *Sky & Telescope*, 6 May 2019, skyandtelescope.org/observing/meet-spica-ear-of-grain/. Accessed 24 July 2023.

"Matariki star facts." Te Papa, www.tepapa.govt.nz/discover-collections/read-watch-play/matariki-maori-new-year/what-and-who-matariki/matariki-star. Accessed 20 August 2024.

NASA: National Aeronautics and Space Administration. "Our Sun: Facts." *NASA Science*, science.nasa.gov/sun/facts/. Accessed 25 June 2024.

National Aeronautics and Space Administration. "Messier 87." *NASA Science*, science.nasa.gov/mission/hubble/science/explore-the-night-sky/hubble-messier-catalog/messier-87/. Accessed 10 June 2024.

National Aeronautics and Space Administration. "Star Basics." *NASA*, science.nasa.gov/universe/stars/. Accessed 7 June 2024.

O'Meara, Stephen James. "Southern Pleiades." *Astronomy Magazine*, 1 January 2024, www.astronomy.com/science/southern-pleiades/. Accessed 6 June 2024.

Open Exoplanet Catalog. "Proxima Centauri d." *Open Exoplanet Catalogue*, www.openexoplanetcatalogue.com/planet/Proxima%20Centauri%20d/. Accessed 6 June 2024.

Plait, Phil. "Aldebaran, seen deeply: A glimpse of the Sun's future." *Bad Astronomy Newsletter*, 9 January 2023, badastronomy.

substack.com/p/aldebaran-seen-deeply-a-glimpse-of. Accessed 14 January 2024.

Ridpath, Ian. *Star Tales*. Universe Books, 1988.

Royal Museums Greenwich. "Stellar jewels: the Pleiades." *Royal Museums Greenwich*, www.rmg.co.uk/stories/topics/what-are-pleiades. Accessed 2 June 2024.

Schilling, Govert. "Mira's Marvelous Tail—Sky & Telescope." *Sky & Telescope*, 15 August 2007, skyandtelescope.org/astronomy-news/miras-marvelous-tail/. Accessed 30 April 2024.

Sol Company. "Achernar / Alpha Eridani." *SolStation.com*, www.solstation.com/x-objects/achernar.htm. Accessed 19 October 2023.

Space.com. "Stars Pronunciation Guide." *Space.com*, 13 December 2006, www.space.com/3250-stars-pronunciation-guide.html. Accessed 10 June 2024.

Stuart, Colin. "Betelgeuse's Great Dimming: The Aftermath—Sky & Telescope." *Sky & Telescope*, 25 August 2022, skyandtelescope.org/astronomy-news/betelgeuses-great-dimming-the-aftermath/. Accessed 26 November 2023.

Tillman, Nola Taylor. "What Is the Most Massive Star?" *Space.com*, 27 July 2018, www.space.com/41313-most-massive-star.html. Accessed 30 April 2024.

Wenz, John. "This is the Smallest Star ever Discovered." *Popular Mechanics*, 11 July 2017, www.popularmechanics.com/space/deep-space/news/a27260/smallest-star-ever-discovered-by-astronomers/. Accessed 12 May 2024.

PHOTO CREDITS

- **Tycho Brahe Portrait:** By R Cooper—Public Domain, commons.wikimedia.org/w/index.php?curid=1178027

- **Earth's City Lights:** Data: By Marc Imhoff/NASA GSFC, Christopher Elvidge/NOAA NGDC; Image: By Craig Mayhew and Robert Simmon/NASA GSFC—visibleearth.nasa.gov/view.php?id=55167 (image link), Public Domain, commons.wikimedia.org/w/index.php?curid=233702

- **Bortle Sky Classification:** By ESO/P. Horálek, M. Wallner—This media was produced by the European Southern Observatory (ESO), under the identifier dark-skies This tag does not indicate the copyright status of the attached work. A normal copyright tag is still required. See Commons: Licensing., CC BY 4.0, commons.wikimedia.org/w/index.php?curid=145950714

- **Messier 87:** By en:NASA, en:STScI, en:WikiSky—wikisky.org/snapshot?img_size=&img_res=&ra=12.5138&de=12.3896&angle=0.0293&projection=tan&rotation=0.0&survey=astrophoto&img_id=905632&width=2160&height=2160&img_borders=&interpolation=bicubic&jpeg_quality=0.8, Public Domain, commons.wikimedia.org/w/index.php?curid=7598267

- **Messier 87 Black Hole:** By NASA/JPL-Caltech/IPAC/Event Horizon Telescope Collaboration—photojournal.jpl.nasa.gov/catalog/PIA23122 (image link), Public Domain, commons.wikimedia.org/w/index.php?curid=78387498

- **Stellar Evolution:** By Courtesy NASA/JPL-Caltech—www.jpl.nasa.gov/infographics/infographic.view.php?id=10737, Public Domain, commons.wikimedia.org/w/index.php?curid=47872358

- **Solar Cycle:** By David Chenette, Joseph B. Gurman, Loren W. Acton—solar.physics.montana.edu/mckenzie/Images/The_Solar_Cycle_XRay_hi.jpg David Chenette at Lockheed-Martin Advanced Technology Center., CC0, commons.wikimedia.org/w/index.php?curid=14896657

- **Layers of the Sun:** By NASA—File is own work, Public Domain, commons.wikimedia.org/w/index.php?curid=115822080

- **Solar Prominence:** By NASA/SDO AIA Team—sdo.gsfc.nasa.gov/firstlight/, Public Domain, commons.wikimedia.org/w/index.php?curid=10102517

- **Coronal Mass Ejection:** By NASA Goddard Space Flight Center from Greenbelt, MD, USA—CME Blow Out, Public Domain, commons.wikimedia.org/w/index.php?curid=51482165

- **Solar Flare:** By NASA Godwdard Space Flight Center from Greenbelt, MD, USA—X Class Solar Flare Sends "Shockwaves" on the Sun, Public Domain, commons.wikimedia.org/w/index.php?curid=51485300

- **Coronal Hole:** By NASA Goddard Laboratory for Atmospheres and Yohkoh Legacy data Archive—NASA Goddard Laboratory for Atmospheres rsd.gsfc.nasa.gov/rsd/images/yohkoh.html rsd.gsfc.nasa.gov/rsd/images/yohkoh_l.gifYohkoh mission of ISAS, Japan. The Yohkoh Soft X-ray telescope was prepared by the Lockheed-Martin Solar and Astrophysics Laboratory, the National Astronomical Observatory of Japan, and the University of Tokyo with the support of NASA and ISAS. ylstone.physics.montana.edu/ylegacy/, Public Domain, commons.wikimedia.org/w/index.php?curid=38853089

- **Solar Wind:** By NASA's Goddard Space Flight Center/Conceptual Image Lab/Adriana Manrique Gutierrez—www.nasa.gov/feature/goddard/2021/switchbacks-science-explaining-parker-solar-probe-s-magnetic-puzzle, Public Domain, commons.wikimedia.org/w/index.php?curid=121284188

- **Sirius and Companion Star:** By NASA, ESA, H. Bond (STScI), and M. Barstow (University of Leicester)—www.spacetelescope.org/images/heic0516a/, Public Domain, commons.wikimedia.org/w/index.php?curid=477445

- **Vega System and Solar System Comparison:** By NASA/JPL-Caltech—www.nasa.gov/mission_pages/spitzer/multimedia/pia16611.html, Public Domain, commons.wikimedia.org/w/index.php?curid=24187596

- **Orpheus and Eurydice:** By Edward Poynter—[1], Public Domain, commons.wikimedia.org/w/index.php?curid=30574721

- **Surface of Betelgeuse:** By ALMA, CC BY 4.0, commons.wikimedia.org/w/index.php?curid=60432670

- **Cowherd and Weaver Girl:** "Rendezvous in the Milky Way." Saint Petersburg: The State Hermitage Museum, Public Domain, commons.wikimedia.org/w/index.php?curid=65025722

- **Antares Orbit and Photosphere:** By NRAO/AUI/NSF, S. Dagnello—public.nrao.edu/news/supergiant-atmosphere-of-antares-revealed-by-radio-telescopes/, CC BY 3.0, commons.wikimedia.org/w/index.php?curid=91534660

- **Fomalhaut Debris Disks:** Image: By NASA, ESA, CSAIMAGE; Processing: By András Gáspár (University of Arizona), Alyssa Pagan (STScI); Science: By András Gáspár (University of Arizona)—webbtelescope.org/contents/media/images/2023/109/01GWWHHHT27VZEQ5D1MK6EHD46, Public Domain, commons.wikimedia.org/w/index.php?curid=131670578

- **Spectral Class:** By Pablo Carlos Budassi—Own work, CC BY-SA 4.0, commons.wikimedia.org/w/index.php?curid=92588077

- **Hipparcos Mission Insignia:** By European Space Agency—Published source: "Hipparcos mission logo" by the European Space AgencyDirect source: Image, hosted by sci.esa.int, Fair use, en.wikipedia.org/w/index.php?curid=52691235

- **GAIA Mission Insignia:** By ESA, CC BY-SA IGO 3.0, CC BY-SA 3.0 igo, commons.wikimedia.org/w/index.php?curid=131420405

- **RMC 136a1:** By ESO/P. Crowther/C.J. Evans—The young cluster RMC 136a, CC BY 4.0, commons.wikimedia.org/w/index.php?curid=11013676

- **Mira A, the star with a "tail":** By NASA—www.galex.caltech.edu/MEDIA/2007-04/images.html, Public Domain, commons.wikimedia.org/w/index.php?curid=3702832

- **Pleiades:** By Jiří Bubeníček—bubenic.cz/fotky/2018/10_hrastice/imgp1869-83.jpg, CC BY-SA 4.0, commons.wikimedia.org/w/index.php?curid=102328551

- **Pleiades in the Caves of Lascaux:** By JoJan—Self-photographed, CC BY 4.0, commons.wikimedia.org/w/index.php?curid=121647870

- **Nebra Sky Disk:** By user "Anagoria"—Self-photographed on 17 August 2012, Public Domain, commons.wikimedia.org/w/index.php?curid=23143773

- **Beehive Star Cluster:** By Fried Lauterbach—Own work, CC BY-SA 4.0, commons.wikimedia.org/w/index.php?curid=115843750

- **Most Brightest Stars of the Night Sky:** By Giorgia Hofer/IAU OAE—This media was produced by the International Astronomical Union (IAU), under the identifier ann22042ac. This tag does not indicate the copyright status of the attached work. A normal copyright tag is still required. See Commons: Licensing., CC BY 4.0, commons.wikimedia.org/w/index.php?curid=133667962

ACKNOWLEDGEMENTS

This book would not have come to fruition without the support of many individuals. I am particularly grateful to Jessica Faroy and Lisa McGuinness for reaching out to me through my YouTube channel, Learn the Sky, and inspiring me to write this book. I also want to thank my editor, Hugo Villabona, for his generosity in granting me numerous extensions and for patiently addressing all my questions. Additionally, I am thankful to Mango Publishing for giving me the opportunity to become a first-time author.

I want to extend my heartfelt thanks to Erin Miller, my illustrator, for her remarkable talent in transforming my thoughts and ideas into stunning works of art. Every time I share an idea, you somehow manage to elevate it a thousand times beyond what I could have envisioned. Collaborating with you as an educator has greatly enhanced my teaching. Nowadays, I cannot think about teaching science without integrating an art project into the mix.

I am deeply grateful to my students and assistants, Abigail Angelisanti and Abigail Reinart, for their unwavering support and encouragement. Their dedication to keeping me focused during the hectic school years has been truly invaluable.

I am incredibly grateful to the teachers, mentors, and colleagues who have guided me throughout my teaching journey. A special thank you to Mr. Talboo, the first teacher who taught me how to share the wonders of the constellations. My time at Hawaii Technology Academy allowed me to explore astronomy through hands-on experience, and I am deeply appreciative of the administrators, staff, and students who supported me along the way. Lastly, to the educators who continue to inspire me—Beth Hartford, Kumu Kaʻea, Kumu Lily, Mr. Lance Leonhardt, Dr. Dru Germanoski, Dr. Lawrence Malinconico, Dr. Varoujan Gorjian, and Dr. Luisa Rebull—your wisdom, support, and passion have profoundly shaped my growth as a teacher, and I am forever grateful.

I am thankful for my students, both those I've taught in person and my online community from YouTube. Your curiosity and questions continually inspire me to keep teaching.

Thank you to all those close friends and family who have supported me and listened to me talk about this big dream of helping others learn about the night sky. Special thanks to Robin Knapp, Tyler Bonham, Brian Holz, Jason Dale, the Flock, my girls Emmie and Isla, and my feline owners Tyke, Satine, Fuji and Albireo.

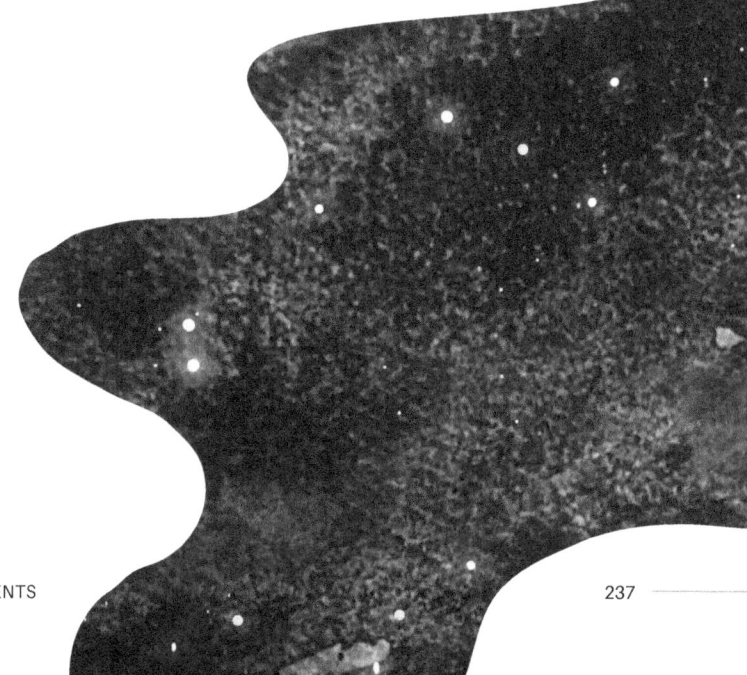

ABOUT THE AUTHOR

Janine Bonham is an experienced educator with expertise in both classroom and online settings, focusing on curriculum development in Earth science, astronomy, biology, and marine science. She earned a bachelor's in environmental biology and geosciences from Lafayette College and a master's in education from Lehigh University. Janine began her teaching career in college as a laboratory assistant in geology and biology labs, helping fellow students. After completing graduate school, she relocated to the Big Island of Hawai'i, where she taught science for twelve years and had the opportunity to explore the islands of Kaua'i, O'ahu, Maui, and Hawai'i. It was during her time in Hawai'i that she deepened her knowledge of the night sky and the patterns of stars.

She is currently a volunteer with the NASA Solar System Ambassadors Program, where she interacts with the public to share information about NASA's space exploration initiatives and discoveries. Her favorite missions include the Parker Solar Probe, the Juno mission, Mars Perseverance, and the James Webb Space Telescope.

Janine is the creator of the YouTube channel *Learn the Sky*, where she teaches viewers about the wonders of astronomy and the night sky. Her goal on this platform is to make astronomy accessible to beginner stargazers.

Outside of academic life, Janine is a mother of two girls and four cats, a scuba diver, a yoga instructor, and has studied hula with Hālau Ka'eaikahelelani. She currently resides in Bethlehem, Pennsylvania, teaching middle school Earth science classes.

ABOUT THE ILLUSTRATOR

Erin Miller is a middle school art educator, creativity coach, and freelance artist based in Pennsylvania. She attended Kutztown University for both her undergraduate and graduate degrees in art education. She is passionate about the value of art and the creative process in understanding, communication, and critical thinking in schools. She primarily works with educators to blend art and the artistic process into all content areas. It is through such collaboration that she was able to work with Janine, and where the ideas for this artwork were born. *Starry Wonders* is the first book she's illustrated.

Erin is a painter who loves to work with mixed media and learn new techniques and processes. She created this artwork with watercolor, pencil, and digital media. She loves all things science and making connections between art and science. When she isn't creating artwork, she loves being outside in nature and spending time with her very large extended family. She has three children, two dogs, and too many cats.

Mango Publishing, established in 2014, publishes an eclectic list of books by diverse authors—both new and established voices—on topics ranging from business, personal growth, women's empowerment, LGBTQ studies, health, and spirituality to history, popular culture, time management, decluttering, lifestyle, mental wellness, aging, and sustainable living. We were named 2019 *and* 2020's #1 fastest growing independent publisher by *Publishers Weekly*. Our success is driven by our main goal, which is to publish high-quality books that will entertain readers as well as make a positive difference in their lives.

Our readers are our most important resource; we value your input, suggestions, and ideas. We'd love to hear from you—after all, we are publishing books for you!

Please stay in touch with us and follow us at:

Facebook: Mango Publishing
Twitter: @MangoPublishing
Instagram: @MangoPublishing
LinkedIn: Mango Publishing
Pinterest: Mango Publishing
Newsletter: mangopublishinggroup.com/newsletter

Join us on Mango's journey to reinvent publishing, one book at a time.

www.ingramcontent.com/pod-product-compliance
Ingram Content Group UK Ltd.
Pitfield, Milton Keynes, MK11 3LW, UK
UKHW020616170625
459667UK00006BA/3